Nitrogen Fixation in Plants

TERTIARY LEVEL BIOLOGY

A series covering selected areas of biology at advanced undergraduate level. While designed specifically for course options at this level within Universities and Polytechnics, the series will be of great value to specialists and research workers in other fields who require a knowledge of the essentials of a subject.

Recent titles in the series:

Locomotion of Animals	Alexander
Animal Energetics	Brafield and Llewellyn
Biology of Reptiles	Spellerberg
Biology of Fishes	Bone and Marshall
Mammal Ecology	Delany
Virology of Flowering Plants	Stevens
Evolutionary Principles	Calow
Saltmarsh Ecology	Long and Mason
Tropical Rain Forest Ecology	Mabberley
Avian Ecology	Perrins and Birkhead
The Lichen-Forming Fungi	Hawksworth and Hill
Plant Molecular Biology	Grierson and Covey
Social Behaviour in Mammals	Poole
Physiological Strategies in Avian Biology	Phillips, Butler and Sharp
An Introduction to Coastal Ecology	Boaden and Seed
Microbial Energetics	Dawes
Molecule, Nerve and Embryo	Ribchester

TERTIARY LEVEL BIOLOGY

Nitrogen Fixation in Plants

R.O.D. DIXON, MSc, PhD
Senior Lecturer in Botany
University of Edinburgh

and

C.T. WHEELER, BSc, PhD
Lecturer in Botany
University of Glasgow

Blackie

Glasgow and London

Published in the USA by
Chapman and Hall
New York

Blackie & Son Limited,
Bishopbriggs, Glasgow G64 2NZ

7 Leicester Place, London WC2H 7BP

Published in the USA by
Chapman and Hall
in association with Methuen, Inc.
29 West 35th Street, New York, NY 10001
© 1986 Blackie & Son Ltd
First published 1986

All rights reserved.
No part of this publication may be reproduced,
stored in a retrieval system, or transmitted,
in any form or by any means,
electronic, mechanical, recording or otherwise,
without prior permission of the publishers.

British Library Cataloguing in Publication Data

Dixon, R.O.D.
  Nitrogen fixation in plants.—Tertiary
level biology
  1. Nitrogen—Fixation 2. Plants
  I. Title II. Wheeler, C.T. III. Series
  581.1'33   QK898.N6

ISBN 0-216-91443-4
ISBN 0-216-91442-6 Pbk

Library of Congress Cataloging-in-Publication Data

Dixon, R.O.D.
  Nitrogen fixation in plants.

  Bibliography: p.
  Includes index.
  1. Nitrogen—Fixation. 2. Plant physiology.
3. Micro-organisms, Nitrogen-fixing. I. Wheeler,
C.T. (Christopher T.) II. Title.
QK898.N6D59 1986   589.9'504133   86-4178
ISBN 0-412-01381-9
ISBN 0-412-01391-6 (pbk.)

Photosetting by Thomson Press (India) Limited, New Delhi.
Printed in Great Britain by Bell and Bain Ltd., Glasgow

Preface

Biological nitrogen fixation has commanded the attention of scientists concerned with plant mineral nutrition for more than 100 years. The importance of this process in managed and natural ecosystems has sustained a substantial research effort which has expanded markedly in recent years, stimulated particularly by increased energy costs of fertilizer production, and ecological concern about excess fertilizer use in agriculture and forestry. The ability to fix nitrogen is exhibited by a broad spectrum of microorganisms such as cyanobacteria, archaebacteria and actinomycetes, each of which occupies an important niche in the nitrogen cycle of particular ecosystems. However, under natural conditions, the highest rates of nitrogen fixation are normally found in symbiotic associations with green plants. These associations are organized within specialized structures which function to the mutual advantage of host and microsymbiont. The *Rhizobium*–legume root nodule is the best known of the symbiotic systems and is of major importance in agriculture. The great advances of recent years in recombinant DNA technology have provided a basis for experimentation which, it is hoped, will lead to improvements in, or the extension of, symbiotic systems and to the ability to fix nitrogen in other organisms. We have therefore emphasized these aspects of the subject, within a broad framework of biochemistry, physiology and genetics.

Nitrogen fixation occurs in courses in botany, microbiology, biochemistry, agriculture and forestry. This book covers the principles of the subject that are common to these disciplines and is not biased towards any one of them. It is written for advanced undergraduates or for postgraduates commencing a study of the subject.

Special thanks are due to various colleagues in Edinburgh and Glasgow who have made helpful comments during the preparation of the manuscript, or who have made available material which is acknowledged in the text. Thanks are also due to Dr C. Jeffree for preparation of many of the figures and to Mr N. Tait and Mr W. Foster for photographic assistance.

RODD
CTW

Contents

Chapter 1. THE ORGANISMS   1

   1.1 Free-living organisms   3
       1.1.1 Anaerobes   3
       1.1.2 Facultative anaerobes   5
   1.2 Aerobes   6
   1.3 Symbiotic bacteria   9
       1.3.1 *Rhizobium*   9
       1.3.2 *Frankia*   11
       1.3.3 Cyanobacteria   13

Chapter 2. THE SYMBIOSES   15

   2.1 Legume root nodules   15
       2.1.1 Recognition   16
       2.1.2 Infection   18
       2.1.3 Nodule development   20
       2.1.4 Determinate nodules   20
       2.1.5 Indeterminate nodules   20
       2.1.6 The formation of bacteroids   22
       2.1.7 Effective nodules   24
       2.1.8 Ineffective nodules   25
   2.2 Actinorhizal root nodules   25
   2.3 The *Rhizobium–Parasponia* association   34
   2.4 Symbioses with cyanobacteria   35
       2.4.1 Lichens   36
       2.4.2 *Azolla*   37
       2.4.3 Cycadaceae   40
       2.4.4 *Gunnera*   42
   2.5 Symbiosis and nitrogen fixation   44

Chapter 3. BIOCHEMISTRY OF NITROGEN FIXATION   45

   3.1 Non-haem iron proteins   47
   3.2 Nitrogenase   49

	3.2.1 Fe protein	49
	3.2.2 MoFe protein	50
	3.2.3 Molybdenum cofactor	50
3.3	The mechanism of nitrogenase	51
	3.3.1 Substrate reduction	52
	3.3.2 Hydrogen evolution	53
	3.3.3 Electron transport to nitrogenase	54
3.4	Hydrogenase	55
3.5	Assay of nitrogen fixation	57
	3.5.1 The Kjeldahl method	57
	3.5.2 The isotopic method	58
	3.5.3 The acetylene reduction method	58

Chapter 4. THE BIOCHEMISTRY AND PHYSIOLOGY OF THE LEGUME ROOT NODULE   61

4.1	Gaseous exchange	61
	4.1.1 Intercellular spaces	64
4.2	Leghaemoglobin	66
	4.2.1 Synthesis of leghaemoglobin	66
	4.2.2 Location of leghaemoglobin	68
	4.2.3 Function of leghaemoglobin	69
4.3	Hydrogen	70
	4.3.1 Hydrogenase	71
4.4	Carbon metabolism	72

Chapter 5. THE BIOCHEMISTRY AND PHYSIOLOGY OF ACTINORHIZAL NODULES   75

5.1	Gaseous diffusion	75
5.2	Hydrogen uptake and evolution	80
5.3	Haemoglobin	81
5.4	Carbon metabolism	81

Chapter 6. NITROGEN ASSIMILATION   87

6.1	Regulation of nitrogenase by combined nitrogen	87
6.2	The primary reactions of ammonium assimilation	92
	6.2.1 Amides	93
	6.2.2 Ureides	95
6.3	Transfer of fixed nitrogen in symbiotic systems	101

Chapter 7. THE GENETICS OF NITROGEN FIXATION   107

7.1	The genetics of the *Klebsiella* nitrogen-fixing system.	107
	7.1.1 The *nif* genes	107
	7.1.2 Regulation of the *nif* genes	110

7.2	The genetics of other nitrogen-fixing systems	114
	7.2.1 *Azotobacter*	115
	7.2.2 *Rhizobium*	116
	7.2.3 Other organisms	117
7.3	Genes concerned with bacterial–plant interaction	118
	7.3.1 Rhizobial symbiotic genes	119
7.4	Host symbiotic genes	120
	7.4.1 Nodulins	121
	7.4.2 Leghaemoglobin	123

Chapter 8. FUTURE PROSPECTS   125

8.1	Field applications of nitrogen fixation	125
8.2	Plant cropping systems	128
8.3	New species	130
8.4	Genetic engineering	131
	8.4.1 Introducing nitrogen-fixing genes into plants	133
	8.4.2 Nodulating new plants	137
8.5	Industrial nitrogen fixation	142

REFERENCES AND FURTHER READING	144
INDEX	153

CHAPTER ONE

THE ORGANISMS

Nitrogen is readily lost from soils due to the processes of nitrification, denitrification and leaching. Unlike other plant nutrients it cannot be replaced by the weathering of rock or soil particles, and has to be replaced by the small amounts of nitrogen present in rain water, by the processes of living organisms, or artificially through fertilizer use. In natural ecosystems, nitrogen supply by the first two methods normally balances losses over a period of time. On soils of very low nitrogen content such as those exposed following glacial retreat, some volcanic soils, forest soils after fire or exposed subsoils, recolonization by pioneer nitrogen fixers will gradually restore fertility. The development of high-yielding crop species and of fertilizer practices designed to maximize crop yield have imposed, on a global scale, additional demands for nitrogen over the past few decades. These demands have been met largely by increased production and utilization of fertilizer nitrogen. The requirements for fertilizer nitrogen have been predicted to increase further in the future, with the global utilization of 5–6 million tonnes used in 1980 increasing by a factor of 2.5 by the year 2000 (Subba Rao, 1980).

Although a large increase in demand can be met by increasing the production capacity, at least in the developed countries, this solution would exact both an economic and ecological price from society. A biological nitrogen fixation uses a great deal of energy, so that greater demand for nitrogen fertilizer will increase the rate at which costly non-renewable sources of energy such as coal and oil are depleted. Fertilizer run-off from agricultural land is already causing concern because of the increasing concentrations of toxic nitrates in drinking-water supplies and the eutrophication of lakes and rivers.

With the current technology for fertilizer production and the inefficient methods employed for fertilizer application, both the economic and ecological costs of fertilizer usage will eventually become prohibitive. In the future, productive agriculture is likely to become much more dependent upon natural processes of nitrogen fixation than at present. This situation, when it arises, will put pressure on the scientists concerned to breed and manipulate plants and microorganisms to fix nitrogen more efficiently. This demand can only be met effectively if we have a full understanding of the biochemistry, physiology and genetics of nitrogen fixation. Such an understanding, coupled with our expanding knowledge of molecular genetics and with advances in the technology of gene transfer, may even permit us to overcome the formidable obstacles involved in engineering non-leguminous crop plants to fix nitrogen. Some of the considerations involved are discussed in Chapter 8.

The only organisms which have the capability to fix nitrogen are prokaryotes and the archaebacteria (see below). Claims made in the past

Table 1.1 Nitrogen-fixing organisms. This table is not intended to be comprehensive but gives a few examples of each type.

1. Free-living nitrogen-fixing organisms
    - (i) Archaebacteria — *Methanosarcina, Methanococcus*
    - (ii) Anaerobes — *Clostridium, Desulphovibrio, Desulfotomaculum*
    - (iii) Facultative anaerobes — *Klebsiella, Erwinia, Enterobacter*
    - (iv) Microaerobes — *Azospirillum, Aquaspirillum, Arthrobacter*
    - (v) Aerobes — *Azotobacter, Beijerinckia, Derxia*
    - (vi) Photosynthetic bacteria — *Rhodopseudomonas, Rhodospirillum, Chromatium*
    - (vii) Cyanobacteria — *Anabaena, Calothrix, Nostoc*

2. Symbiotic systems
    - (i) *Rhizobium*–legume associations
        - (a) Fast growers with — *Pisum, Trifolium, Vicia*
        - (b) Slow growers with — *Arachis, Glycine, Vigna*
    - (ii) *Rhizobium*–non-legume associations
        - Slow growers with — *Parasponia*
    - (iii) *Frankia* actinorhizal associations with — *Alnus, Casuarina, Myrica*
    - (iv) Cyanobacterial associations
        - (a) With angiosperms — *Gunnera*
        - (b) With gymnosperms — *Agathis, Cycas, Macrozamia*
        - (c) With pteridophytes — *Azolla*
        - (d) With bryophytes — *Anthoceros, Blasia, Caricula*
        - (e) With lichens — *Collema, Lichina, Peltigera*

that some eukaryotic systems, particularly mycorrhizas, can fix nitrogen have always been refuted when more rigorous analysis has been done.

The ability to fix nitrogen is present in a wide range of organisms as the representatives listed in Table 1.1 show, although only a very small proportion of species is able to do so; about 87 species in 2 genera of archaebacteria, 38 genera of bacteria, and 20 genera of cyanobacteria have been identified as diazotrophs, or organisms that can fix nitrogen. The wide variety of diazotrophs ensures that most ecological niches will contain one or two representatives and that lost nitrogen can be replenished.

The nitrogenase enzymes are irreversibly inactivated by oxygen and the process of nitrogen fixation uses a large amount of energy. This is discussed in more detail in Chapters 3 and 8. These requirements for anaerobiosis and energy have to be met by the organisms and determine the conditions under which nitrogen fixation can take place. The high energy requirement is the reason for the predominance of the symbiotic and photosynthetic systems: the *Rhizobium*–legume association and the cyanobacteria.

The pie chart (Figure 1.1) shows that the symbiotic systems and the cyanobacteria fix a large proportion of the nitrogen. A large part of the fixation in grasslands will be by forage legumes such as clovers, and that in woodlands will be by leguminous trees and the woody actinorhizal plants such as alder. (The term 'actinorhizal' was suggested in 1979, to distinguish *Frankia* nodulated non-legumes from the root nodule association between *Rhizobium* and *Parasponia*, a member of the Ulmaceae, and will be used throughout this text).

The non-legume fixation attributed to agriculture is mainly composed of fixation by cyanobacteria in rice paddy fields, either free or in association with *Azolla*. The contribution of fixed nitrogen by the free-living heterotrophic diazotrophs will be important only in natural ecosystems where most of the plant nutrients are in the biomass and the turnover of nitrogen will be very slow.

1.1 Free-living organisms

1.1.1 *Anaerobes*

The archaebacteria. The organisms that belong to this group are diverse but differ from both eukaryotes and prokaryotes and have therefore been placed in a separate kingdom, the archaebacteria. The prokaryotes are now known as the eubacteria. The archaebacteria are all strict anaerobes and divide into three main groups based on their metabolism: the methanogens,

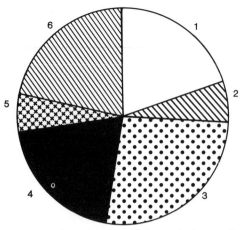

Figure 1.1 Pie chart to show the proportion of the total nitrogen fixed within different habitats.

	Metric tons per year (× 10⁶)	Percentage
A. Arable agriculture	**44**	**25**
1. Grain legumes	35	20
2. Non-legumes	9	5
B. Non-arable agriculture and forestry	**85**	**48**
3. Permanent pasture and grassland	45	26
4. Forest and woodland	40	22
C. Non-agricultural habitats	**46**	**26**
5. Land	10	6
6. Sea	36	20

those bacteria that synthesize and evolve methane as a result of their metabolism; extreme halophiles, bacteria that live in habitats with a very high salt concentration; and thermoacidophiles, bacteria that can thrive in conditions of high temperature and acidity.

The archaebacteria differ from the eubacteria in a number of ways. These include the structure and composition of RNA molecules and differences in cell wall composition; the archaebacteria have no muramic acid in their cell walls although this molecule is present in all eubacterial cell walls. The membrane lipids also differ as the fatty acids are not esterified to glycerol but bound to glycerol by ether linkages. The long-chain fatty acids have branching methyl groups at regular intervals along their length. An account of these interesting organisms is given by Woese (1981).

Recently two species of archaebacteria have been shown to fix nitrogen; *Methanosarcina barkeri* and *Methanococcus thermolithotropicus*. Both are methanogenic bacteria. *Methanococcus thermolithotropicus* is a thermophile and can fix nitrogen at high temperatures (64°C), whereas the eubacteria have a maximum temperature for nitrogen fixation of 35–40°C. As the genes which code for the nitrogenase proteins, *nif* KDH, are homologous with the *nif* KDH genes of *Klebsiella* it is not likely that the nitrogenase enzymes are different, but it is probable that these organisms will have different control genes to enable them to fix nitrogen at these high temperatures (Postgate, 1984).

Clostridium. The first organism which was shown to be capable of nitrogen fixation was *Clostridium pasteurianum*. This was one of the many findings the famous Russian bacteriologist, Winogradsky, made during a distinguished career. This organism is Gram positive and is a strict anaerobe. It lives in the soil and is most frequent where there is adequate organic matter. Even in the upper parts of the soil there will be anaerobic centres to the soil crumbs. The numbers of clostridia are high in the rhizosphere, and although it might seem strange to have an anaerobe living next to plant roots which require adequate oxygen, the oxygen concentration will be lowered in some places because nearby aerobic microorganisms use it for their respiration. Although nitrogen fixation is an anaerobic process, only two other genera of prokaryotes contain diazotrophic strains: these are the sulphur-reducing bacteria, *Desulphovibrio* and *Desulfotomaculum*.

1.1.2 Facultative anaerobes

Klebsiella. This genus and several others contain species of facultative anaerobic bacteria which can fix nitrogen. They normally fix nitrogen only under anaerobic conditions as they have no means of protecting the nitrogenase from oxygen. In the laboratory *Klebsiella* has been shown to reduce nitrogen under microaerobic conditions (Hill, 1976). Indeed, because of the more efficient aerobic metabolism, at low oxygen concentrations nitrogen reduction can proceed at up to twice the rate of anaerobic reduction. Derepression, or the switching on, of nitrogenase is more oxygen-sensitive than the reduction reaction itself. Thus although 80 nM O_2 partially repressed nitrogenase synthesis, nitrogenase activity, once it had been switched on anaerobically, was maintained at up to 550 nM O_2, 0.22% of atmospheric oxygen (Hill *et al.*, 1984). There is thus a certain amount of leeway if a small amount of oxygen should diffuse into the anaerobic sites in which *Klebsiella* is fixing nitrogen.

Facultative anaerobes have the flexibility to grow in both aerobic and anaerobic environments. When growing aerobically there is a wide range of modes of existence among these bacteria. Some of them, such as *Erwinia herbicola* and *Klebsiella pneumoniae*, may exist as pathogens. When the bacteria inhabit an aerobic environment, mutants that cannot fix nitrogen will not be at a disadvantage. Mutations in the genes that control nitrogen fixation will not be eliminated and can accumulate. It is thus not surprising to find that about 50% of the strains of *K. pneumoniae* do not fix nitrogen. There is evidence, based on DNA homology, that some of these non-fixing strains contain some of the nitrogen-fixing genes.

Klebsiella pneumoniae is a very widespread organism and can grow in the rhizosphere of plants, in the soil and on leaf surfaces as well as being an animal pathogen, although in the latter case it occurs only as a secondary infection after a more virulent pathogen has entered and thus lowered the host's resistance to infection.

1.2 Aerobes

Azotobacter. Members of this genus are strict aerobes: oxygen is required for metabolism, and also to fix nitrogen. Nitrogen fixation therefore occurs in an aerobic environment and there must be a mechanism to prevent the access of oxygen to the oxygen-sensitive proteins. *Azotobacter* has a very high rate of respiration, and when the organism is deprived of respirable substrate, as when it is grown on a medium low in carbon, the nitrogenase is more susceptible to oxygen.

These observations led to the concept of respiratory protection. The organism is able to respire sufficiently rapidly to prevent oxygen penetrating to the nitrogenase proteins. Evidence for this is that nitrogenase is inhibited if the oxygen concentration is raised to a level at which the bacteria can no longer respire quickly enough to prevent the entry of oxygen (Dalton and Postgate, 1968). When bacteria are deprived of phosphate, which limits the respiration rate, the same effect of nitrogenase inhibition is obtained. Under conditions in which oxygen concentrations are high, i.e. approaching normal atmospheric concentration, about 240 μM in solution, the respiration is partially uncoupled from the production of ATP, thus permitting the high respiration rates. Nitrogen fixation is inhibited when the organism is suddenly exposed to an oxygen concentration higher than that in which it has been grown. If the oxygen concentration is then lowered again, nitrogen reduction quickly recommences before there has been time for the synthesis of new nitrogenase proteins. This 'switch off, switch on' effect shows that during the period of high oxygen concentration the nitrogenase proteins were not inactivated but protected, and were unable to operate in the protected state. Further evidence for this protected state is that crude extracts of *Azotobacter* have little oxygen sensitivity, but when the components are purified they are as sensitive to oxygen as those in any other bacterium.

On this evidence it was proposed that the proteins changed their shape, or conformation, when in contact with oxygen and that this protected the active sites of the enzymes from oxidation. This then became known as 'conformational protection'. However, it is now known that the proteins do not change shape but combine with the oxidized form of a third protein known as FeS II. The resulting protein complex protects the active sites of the enzymes. When the oxygen concentration is lowered, FeS II becomes reduced and dissociates from the complex and the enzymes become active again. Figure 1.2 illustrates part of the evidence for this. When the two nitrogenase proteins have been combined with differing amounts of FeS II, protection is increased as the concentration of FeS II is increased. The complexed proteins sedimented together in the ultracentrifuge and on analysis it was found that FeS II was present in equimolar amounts with nitrogenase proteins (Robson, 1979).

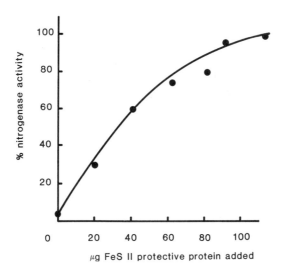

Figure 1.2 Oxygen protection of nitrogenase by FeS II protein. Purified Fe and FeMo proteins were mixed with different amounts of the protective protein, FeS II, and then exposed to oxygen for two minutes. The activity that remained was then assayed by acetylene reduction. Data from R.L. Robson (1979).

There are thus two mechanisms of oxygen protection in *Azotobacter*: respiratory protection and protection with the FeS II complex. Respiratory protection by its nature uses a lot of energy, as carbon substrates have to be respired for protection. *Azotobacter* therefore grows better at oxygen concentrations lower than atmospheric when fixing nitrogen. Where there is sufficient substrate it is likely that the oxygen concentration will be lower than atmospheric because of the respiration of both *Azotobacter* and of other microorganisms in the vicinity.

Azotobacter has an uptake hydrogenase that recycles hydrogen produced from nitrogenase and thus saves some energy. This hydrogenase is synthesized under conditions in which the nitrogenase operates, and like nitrogenase it is repressed by ammonia (Hyndman and Burris, 1953). This is discussed more fully in Chapter 4. *Azotobacter* lives in the rhizosphere of plants as it needs a lot of energy. There is, however, a specific association of *Azotobacter* and the tropical grass *Paspalum notatum*. The bacteria that form this association have been put in a separate species *A. paspali* (Dobereiner *et al.*, 1972).

Derxia. This genus is related to *Azotobacter* but is more oxygen-sensitive and is unable to fix nitrogen at the oxygen concentration found in air. It secretes large amounts of extracellular polysaccharide, which accounts for its specific name *D. gummosa*. When large colonies are formed, the respiration of the cells in the outer part of the colony reduces the amount of oxygen diffusing to the centre of the colony. This makes the centre less aerobic and enables the bacteria there to fix nitrogen. It is a tropical genus, and its oxygen sensitivity is reflected in the fact that it is most common in flooded soils. *Derxia*, like *Azotobacter*, has an uptake hydrogenase, can use hydrogen as the sole energy source, and also grow autotrophically with H_2, O_2 and CO_2. In flooded habitats it may well receive hydrogen from the products of anaerobic organisms.

Azospirillum. This genus of spirally curved bacteria is interesting as its members not only live in the rhizosphere of grasses but can also enter the root cortex. *Azospirillum* is oxygen-sensitive and can fix nitrogen only at low oxygen concentrations. It is a tropical bacterium and has a high optimum temperature so that it does not occur to any great extent in temperate latitudes. It has a wide host range and some work has been done to see if it will provide any agricultural benefit for cereal crops. Results so far have been disappointing, and it has been suggested that those beneficial effects which have been found are due to the production of growth-regulating substances rather than to the amounts of nitrogen that have been fixed. Similarly, other nitrogen-fixing bacteria, when added to crops, have sometimes stimulated growth, and such stimulation has also been ascribed to the production of plant growth-regulating substances (Subba Rao, 1980).

1.3 Symbiotic bacteria

1.3.1 *Rhizobium*

The pie chart (Figure 1.1) shows that grain legumes themselves contribute 20% of nitrogen fixed, and considering that most of the fixation in grasslands will be by forage legumes and a proportion of the fixation in forest and woodlands will be by tree legumes, we can arrive at the conservative estimate that 50% of natural nitrogen fixation is accomplished by the *Rhizobium*–legume association. Of the nitrogen fixation of agricultural value and hence of importance to man, the proportion rises to over 70%.

The rhizobia are soil organisms that inhabit the rhizosphere of legumes and other plants. They are a rather more diverse group of organisms than might be supposed, but are united by their ability to produce nodules on legumes. A proper understanding of the *Rhizobium*–legume association requires a knowledge of the relationship of rhizobia one to another and to other members of the Rhizobiaceae. There are two main types of rhizobia, distinguished by a number of differences but referred to according to their growth rate on laboratory media as 'fast growers' and 'slow growers'*. The division of *Rhizobium* into species is based on the interaction with plants; those bacteria which nodulate clovers, for instance, are put in *R. trifolii* and those that nodulate peas and vetches are put in *R. leguminosarum*. Although this classification is still used, it has been realized for a long time that it is not satisfactory. In the slow-growing rhizobia there is a group which has such a wide specificity that it has not been assigned to a species, but is called the cowpea group after one of the many hosts. Since the classification was first accepted, more and more exceptions have been found with regard to species host specificity. Some strains can be put into more than one species but generally retain the name from the host from which they were first isolated.

Agrobacterium is a member of the Rhizobiaceae and this genus also interacts with plants. *A. tumefaciens*, *A. rubi* and *A. rhizogenes* are pathogens on dicotyledons; *A. tumefaciens* and *A. rubi* form crown galls.

*Currently the tendency is to place the slow-growing rhizobia in a separate genus, *Bradyrhizobium*.

Table 1.2 Some diagnostic tests in the Rhizobiaceae

	Agrobacterium	Rhizobium Fast growers	Rhizobium Slow growers
Form	Gram⁻ rods	Gram⁻ rods	Gram⁻ rods
Flagellae	peritrichous	peritrichous	polar/subpolar
Growth	fast	fast	slow/variable
Acid/alkali	acid	acid	alkali
%GC	59–63.5	59.5–63	59.5–65.5
Serum zone in litmus milk	+	+	−

In these tests the fast-growing rhizobia are more similar to *Agrobacterium* than they are to the slow-growing rhizobia.

A. rhizogenes causes proliferation of adventitious roots or hairy root disease. Table 1.2 shows the results of some diagnostic tests from which it can be seen that *Agrobacterium* is more similar to the fast-growing rhizobia than the fast-growing rhizobia are to the slow growers. Of course one requires more evidence than is given in this table to make taxonomic judgements. A number of studies have been done using many more characters. Studies using the techniques of numerical taxonomy, where every character is given equal weighting, have used over 100 different characters in their analysis (Moffet and Colwell, 1968). It is now generally agreed that the two types of rhizobia are distinct from one another and sufficiently distinct to put them into separate genera. The closeness of *Agrobacterium* to the fast-growing rhizobia has also led to the suggestion that they be included in the same genus, *Rhizobium*.

The placing of *Agrobacterium* with *Rhizobium* may seem odd because they have such different effects on plants. However, more recent work has shown that in both *Agrobacterium* and *Rhizobium* the characters concerned with the plant interaction are coded for on a plasmid (circular non-chromosomal DNA). It was first found that the pathogenic properties of *A. tumefaciens* correlated with the possession of a plasmid and then that the removal of the plasmid removed the pathogenic properties. It is now known that *Agrobacterium* transfers part of the plasmid DNA to the plant where it becomes incorporated into the plant's chromosomes. Subsequently it was discovered that the nodulating properties of the fast-growing strains of *Rhizobium* were coded for by plasmid genes, as were also the nitrogen-fixing enzymes. Plasmids bearing the nodulating and nitrogen-fixing genes have not so far been found in the slow-growing rhizobia. In this group the genes are probably therefore on the chromosome. Plasmids will be discussed in more detail in Chapter 8.

As well as forming root nodules, for which they are recognized, some strains are capable of forming nodules on stems, although it is root primordia on the stems that become infected. *Sesbania rostrata* and *Aeschynomene afraspera* are two species that can be so nodulated. Some of the strains of *Rhizobium* which form stem nodules are able to fix sufficient nitrogen to support growth in the free-living state. In order to do this, however, the oxygen concentration must be low, as they have no means of oxygen protection. The slow-growing rhizobia in general are able to fix nitrogen in culture but are unable to fix sufficient nitrogen to grow. The fast-growing rhizobia have not so far been found to fix nitrogen outside the root nodule.

Considerable interest has been aroused by the finding that some strains of *Rhizobium* can form effective nodules with some tropical non-legumes, species of *Parasponia*, a genus in the Ulmaceae (Akkermans *et al.*, 1978). The interest is not because of the economic importance of these plants but because of the questions raised with regard to the evolution of the *Rhizobium*–plant relationship.

1.3.2 *Frankia*

The members of this genus are actinomycetes: most of these bacteria at some time in their life cycle have a filamentous habit which often superficially bears some morphological resemblance to the fungi. They are, however, prokaryotes with hyphae of smaller dimension–in *Frankia* typically less than 2 μm diameter–than fungi. This genus has been defined in a similar manner to *Rhizobium* as comprising those actinomycetes that can form root nodules on plants. Originally it was proposed that the species name should be related to the host plant in a similar fashion to that in rhizobia. However, the promiscuity of some strains which nodulate more than one plant genus, together with the influence which the host plant can have on endophyte morphology *in vivo*, have been major factors in the decision to defer speciation of the genus until more information is available.

The many inconclusive attempts to isolate and to grow *Frankia* in culture before 1978 led to suggestions that this organism might be an obligate symbiont of actinorhizal plants, even though indications of extranodular growth of *Frankia* had been obtained from the increased capacity of soils to nodulate the host plant after inoculation and incubation with crushed nodule suspensions. Such speculation was dispelled in 1978, when Callaham, del Tredici and Torrey isolated *Frankia* from *Comptonia peregrina* nodules and were able to satisfy Koch's postulates with this isolate. With hindsight, *Frankia* is not an especially difficult organism to

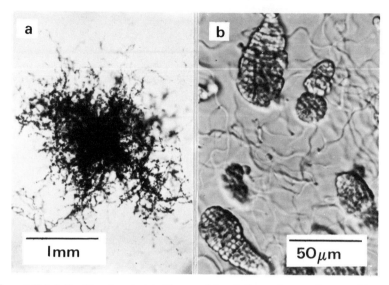

Figure 1.3(*a*) A *Frankia* colony isolated from nodules of *Alnus rubra* growing near Glasgow and cultured on agar supplemented with propionate, hydrolysed casein, biotin and minerals. (*b*) *Frankia* isolated from *Alnus rubra* showing peanut-shaped sporangia borne on colonies of this strain. Photographs, J.E. Hooker and C.T. Wheeler.

isolate, and many strains can be grown on media used traditionally for the culture of soil actinomycetes such as propionate medium or glucose asparagine agar (Figure 1.3*a,b*). The main problem encountered is the slow growth of isolates—colonies may take from a couple of weeks to several months to appear on isolation plates. Special methods must be adopted therefore to prevent isolation plates being overrun by contaminant microorganisms before *Frankia* colonies develop. Such methods either employ special sterilizing techniques to destroy nodule contaminants, or may use techniques to separate *Frankia* from other microbial contaminants by microfiltration or density gradient centrifugation of nodule homogenates.

An important feature of *Frankia* is that many strains can fix nitrogen at normal oxygen concentrations at rates sufficient to support growth in culture (Tjepkema *et al.*, 1981). Nitrogen fixation is accompanied by the development of terminal swellings on branch hyphae. These structures are known as vesicles, in which nitrogenase is believed to be localized and which have an outer wall of laminated structure which may help to protect nitrogenase by restricting oxygen diffusion (Figure 5.1). Nitrogen fixation in such cultures is inhibited by the addition of combined nitrogen.

1.3.3 Cyanobacteria

Some photosynthetic bacteria such as *Rhodospirillum* and *Rhodopseudomonas* can fix nitrogen but are not quantitatively as important as are the cyanobacteria (blue-green algae). They are important both in the free-living form and in symbiosis. They can grow in both salt and fresh water according to species, and their energy requirements can be satisfied by photosynthesis. Cyanobacteria can fix nitrogen in both aerobic and anaerobic conditions although aerobic nitrogen fixation in filamentous species is confined to those species which possess special cells, heterocysts, for the purpose.

Heterocysts are cells which are larger than vegetative cells, have thicker cell walls and are thus easily recognized (Figure 1.4). They possess less organized and fewer thylakoids and less photosynthetic pigment than do the vegetative cells. The explanation for this is that they lack photosystem II which produces oxygen which would inhibit nitrogenase activity. This means that light energy can only be channelled into the production of ATP by cyclic photophosphorylation and no reductant is formed. Photosynthate is imported from the vegetative cells and metabolized in order to provide reductant for nitrogen fixation. As well as eliminating the

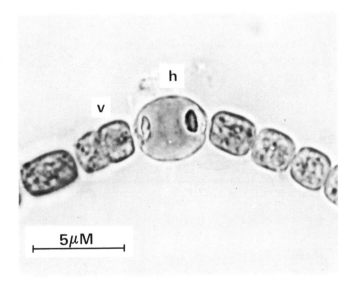

Figure 1.4 A filament of *Anabaena azollae* with a heterocyst, *h*, and vegetative cells, *v*.

evolution of oxygen from its photosynthetic processes the heterocysts have to limit the diffusion of oxygen into the cell from the outside. It has been proposed that the thick cell wall will contribute a resistance to the diffusion of oxygen.

Some unicellular cyanobacteria such as *Gloeotheca* can fix nitrogen aerobically. As they are unicellular, the nitrogenase cannot be separated spatially from photosynthesis but it is separated temporally. Most of the nitrogen fixation is done in the dark at the expense of carbohydrate reserves built up by photosynthesis during the day. Some species of cyanobacteria that do not possess heterocysts can fix nitrogen under anaerobic conditions, as can also the heterocystous species.

The cyanobacteria are obviously most significant in aquatic environments, and are of importance agriculturally in the rice fields. They fix nitrogen when the fields are flooded and rot down and release nitrogen to the rice as the fields dry out. It has been shown that the proper selection of species and strain can influence the yield of rice.

The theme of this chapter has been the variety of free-living organisms that fix atmospheric nitrogen and the adaptations, both morphological and physiological, that have to be made in order to enable some of them to do so under aerobic conditions. The enzyme itself must be anaerobic and it must also have a good supply of energy as ATP and reductant. The energy requirement restricts the amount of nitrogen that can be fixed in natural environments. The supply of energy is more plentiful when the organism is present in a symbiotic association and supplied by the host. Nitrogen-fixing symbioses are discussed in the next chapter.

CHAPTER TWO

THE SYMBIOSES

In the first chapter the heterocyst, a structural modification that helps to enable nitrogen fixation to take place in an aerobic environment, was described. Physiological modifications such as lack of photosystem II in the heterocyst and respiratory protection in *Azotobacter* were also discussed. In the symbioses, modifications to the host enable the partners to form the symbiosis, and the structure and physiology of both host and microorganism are adapted for the aerobic fixation of nitrogen.

This chapter is concerned with the formation and structure of some nitrogen-fixing symbiotic systems. Discussion of their physiology is deferred until Chapters 4 and 5, after the biochemistry of nitrogen fixation has been dealt with in Chapter 3.

2.1 Legume root nodules

The Leguminosae is a large family of about 750 genera and 20 000 species. However, not all species or genera associate with *Rhizobium* to form root nodules. The legumes are important in agriculture not only for the fact that they fix nitrogen and can thus conserve nitrogen fertilizers, but also because the plants and seeds are high in protein. The seeds of grain legumes are important nutritionally for humans, and the forage legumes (clovers, lucerne or alfalfa and vetches) also supply a high-protein diet to livestock. Legumes also have ecological significance as they provide nitrogen inflow to natural habitats such as tropical forests. Some tropical legume trees are grown for their wood and the nitrogen-enriched soil is exploited when the wood is removed. There are several other uses of legumes, but even so only a tiny fraction of the total number of species in this family is exploited by man.

Because there is such a wide range of plants in this family, nodule form and modes of infection will differ in detail according to the host species and rhizobial strain. The discussion here is limited to a few examples in order to present more clearly the general principles.

2.1.1 Recognition

Plant defence mechanisms normally prevent the entry of microorganisms that may cause disease, so that there must therefore be some means of mutual recognition between the symbiotic partners to enable infection to occur and the association to develop. However, when it is growing with adequate mineral nitrogen the symbiosis is not an advantage to the plant because of the high cost to the plant of carbon substrates required to develop and maintain the symbiotic condition. Legumes thus resist infection and nodulation is inhibited if the nitrogen content of the soil is high. When the soil is depleted of nitrogen, rhizobia are recognized and infection occurs. The recognition process is thus controlled by the nitrogen status of the environment through its interaction with the plant. Rhizobia have a restricted host range; each strain can only infect and nodulate a limited number of legume species, so that there must be a high degree of specificity in the recognition process.

The basis of recognition must be the molecules on the surfaces of the partners, and these must be sufficiently diverse if specificity is to be attained. Suitable molecules are proteins and polysaccharides. The range of sugars in the surface polysaccharides of both plant and bacterium is wide enough to provide sufficient diversity. If the sugars are the recognition signals, then there must be a mechanism to distinguish specific sugars and to mediate the recognition response between plant and bacterium. Lectins, proteins which have several sites which will bind sugar molecules, could act in this way, since different lectins show specificity for different sugars. Lectins have several binding sites for sugars and can thus attach the bacterium to the root surface by binding specific sugars on the plant surface and the bacterial surface through binding sites on the same lectin molecule (Figure 2.1). There is now a body of evidence that is consistent with the binding of sugars by lectins as being the recognition process between legume and *Rhizobium*.

The first experiments that suggested lectins might be recognition proteins in the legume–*Rhizobium* association exploited the property of some lectins to bind to and agglutinate red blood cells. Such lectins are known as phytohaemagglutinins and have been used for blood typing, as sugars on red cell membranes differ according to the blood group. They are often found in quantity in the seed. Hamblin and Kent (1973) first showed that

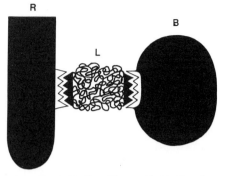

Figure 2.1 The mechanism of lectin binding. The specific binding sites on the lectin attach to sugars on the *Rhizobium* and root hair surfaces. *L*, lectin; *R*, root; *B*, bacterium.

the phytohaemagglutinating lectin of *Phaseolus vulgaris* bound to *Rhizobium* cells and that red blood cells bound to the same portions of the root's surface to which *Rhizobium* binds. Subsequently, Bohlool and Schmidt (1974) attached a fluorescent marker to soya bean lectin so that it could be identified visually. They detected binding to 22 out of 25 strains of *R. japonicum* that could nodulate soya bean, but there was no binding to any of the 23 strains representative of those rhizobia that could not nodulate soya bean. Although the correlation of lectin binding with nodulating capability is not absolute, this, together with the results of Hamblin and Kent, provided a stimulus for the large amount of work that has since been done on this topic.

Experiments by Dazzo and his colleagues illustrate the kind of evidence that supports a role for lectins in legume–*Rhizobium* recognition (Dazzo and Elkan, 1979). They raised an antibody to clover lectin, and attached a fluorescent marker to it. The amount of lectin could then be determined by measuring the fluorescence of lectin–antibody complexes. By using radioactively labelled antibody, the amount of lectin on the root surface could be measured. Nitrate was used to obtain various degrees of inhibition of nodulation. Nitrate also inhibited *Rhizobium* attachment, and it could be shown that the degree of rhizobial attachment was proportional to the amount of lectin present on the root surface. These experimental results suggest that the inhibitory effect of nitrate on nodulation could be due to the inhibition of lectin production on the root surface which then decreases *Rhizobium* binding, the first stage of the infection process.

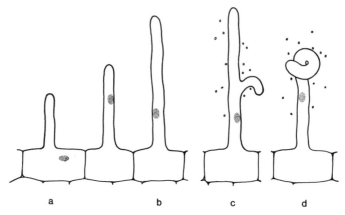

Figure 2.2 Types of root hair deformation: *a*, normal root hair; *b*, root hair elongation; *c*, root hair branching; *d*, root hair curling.

2.1.2 Infection

After recognition, the next stage of the infection process is the morphological alteration of the root hairs. These become elongated, curled and branched (Figure 2.2). As morphological changes on the root hairs occur when the bacteria are outside the host, there must be an interchange of small diffusible molecules between the host and the bacterium. A polysaccharide fraction, isolated from rhizobial growth medium, has been shown to cause root hairs to curl moderately and also to branch. However, the presence of active rhizobia is required to induce the strong root-hair curling that takes place on infection, so that some other active substance may also be necessary.

The mechanism of infection is a complicated process which, because it is difficult to observe and also difficult to manipulate experimentally, is not well understood. Although a curled root hair is not essential for infection, it is thought to facilitate infection as the bacteria become enclosed by the root hair walls. This produces a closed environment in which enzymes involved in root hair penetration are prevented from diffusing away. The root hair cell wall is dissolved away at a localized site and the bacterium can then enter (Callaham and Torrey, 1981). On entry, a new wall is formed, cutting the bacterium off from the host cell's contents. At its edge this wall is connected to the root hair wall and it thus appears that the root hair cell wall invaginates at this point (Figure 2.3). The bacteria divide in the enclosed space and the wall enclosing them extends to grow as a tube

THE SYMBIOSES

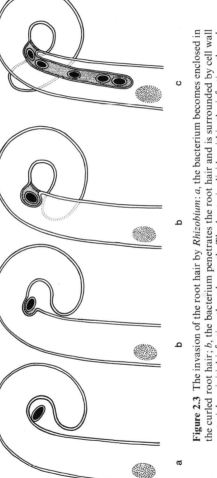

Figure 2.3 The invasion of the root hair by *Rhizobium*: *a*, the bacterium becomes enclosed in the curled root hair; *b*, the bacterium penetrates the root hair and is surrounded by cell wall material; *c*, initial infection thread growth. The bacteria divide within the infection thread.

towards the base of the root hair cell. This tube is termed the infection thread. Although they are within the root, as the infection thread wall is plant cell wall and it, in turn, is surrounded by the plasmalemma, the bacteria are extracellular.

2.1.3 Nodule development

The infection thread continues to grow beyond the root hair cell and penetrates the cortex of the root. When the infection thread has penetrated some way into the cortex, the cells in the inner cortical region are stimulated to divide. The dividing cells are tetraploid, and it was once thought that it was only tetraploid cells that were caused to come into division. However, it is now known that diploid cells divide and become tetraploid by duplication of the chromosome material (Libbenga and Haaker, 1973). The initial divisions are, more often than not, opposite the xylem poles and the nodules emerge in the same position as the lateral roots. They are not, however, modified lateral roots as they are initiated in the cortex and not in the pericycle as are true lateral roots.

At this stage nodule development may vary depending upon the host. The main types of nodule show either determinate growth, as found on soya beans, or indeterminate growth, found on peas. When once formed, the determinate nodules are not capable of further growth by cell division, whereas the indeterminate types maintain a meristem and are capable of continued growth over long periods.

2.1.4 Determinate nodules

In these nodules the bacteria are released into dividing cells, as a result of which the bacteria are distributed through the central cell mass of the nodule. The cells finally cease to divide, so that further increase in nodule size occurs only through cell enlargement. The cortex contains vascular strands that connect with the root vascular system. Within the cortex is an 'endodermis' layer with cells which have suberized walls, which may or may not have Casparian bands. This layer has no intercellular spaces and thus presents a diffusion barrier to gases passing into and out of the nodule.

2.1.5 Indeterminate nodules

Indeterminate nodules differ in a number of ways from determinate nodules. The bacteria are released from the infection threads into newly-divided cells that do not divide again. This means that the bacteria must be distributed within the nodule by the infection thread which continues to grow and branch so that cells behind the dividing cells become infected. The cells that are dividing form a meristem at the apex of the nodule and the growth that results gives the nodule its elongate rather than spherical shape. As the nodules enlarge still further, the meristem itself may divide and give rise to branched nodules.

The continued growth of the nodule means that the nodule can be divided into zones of different age with the oldest cells nearest to the axis of the root (Figure 2.4). At the apex of the nodule is the meristem which produces some diploid cells on the outer side, giving rise to the nodule cortex and tetraploid cells towards the inside. Next is a zone of differentiating tetraploid cells which become enlarged following infection with rhizobia, and this is followed by a region of mature cells which contain bacteroids and in which nitrogen fixation takes place. The pink colour of this region is due to the leghaemoglobin which is contained within the

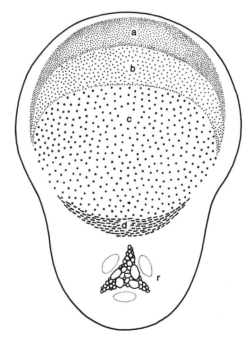

Figure 2.4 Diagram of an indeterminate nodule to show the different zones: a, meristematic zone; b, zone of cell infection and differentiation; c, active nitrogen-fixing tissue; d, senescent cells; r, root.

infected cells. Finally, if the nodule is old enough, there is a region of senescing nodule cells. These can be distinguished as the leghaemoglobin breaks down to form a green pigment.

The vascular bundles are in the cortex, but are open at the ends where they continue to differentiate, unlike the closed vascular system of the determinate nodules. The pericycle of these indeterminate nodules often contains transfer cells (Pate and Gunning, 1969) which are often found in plants at sites where a large interchange of substances takes place between cells. They have deeply infolding walls which increase the surface area of the plasmalemma and thus speed up the process of diffusion by providing a greater area over which solutes can diffuse. Transfer cells are not always present in indeterminate nodules and are not found in determinate nodules. The cortex also contains a layer of cells lacking intercellular spaces which is continuous with the parent root. The cell walls of this 'nodule endodermis' layer are not always suberized, nor do they have a Casparian strip, but they form a diffusion barrier to gases.

The infected cells of both types of nodule are enlarged and are tetraploid, or may even be of higher ploidy (Mitchell, 1965). The nodule cortex, however, remains diploid. As only polyploid cells are infected the extra DNA must have some significance, but the reason for it is at the moment unknown.

2.1.6 *The formation of bacteroids*

The bacteria are released from the infection thread at a point where it is free of host cell wall. The bacteria migrate first from the centre of the infection thread, where they are encased in extracellular polysaccharides, to the periphery of the thread, and are released into the plant cell cytoplasm and surrounded by the plasmalemma by a process of endocytosis (Figure 2.5). Thus even within the host cell they are separated from the contents of the host cell (Dixon, 1967). This membrane is now termed the peribacteroid membrane. The contents of the host cell are seen to be undamaged when they contain bacteria. The bacteria divide and in some cases, such as in pea and clover, the peribacteroid membrane divides also so that each bacterium is surrounded by its own membrane envelope. In other cases, such as soya bean, the peribacteroid membrane does not divide as frequently as the bacteria and as a result each membrane envelope may contain several bacteria. When the cytoplasm is almost full of bacteria, the bacteria enlarge and change shape and become bacteroids. The bacteroids of soya bean differ little in morphology from the bacteria although they may be slightly

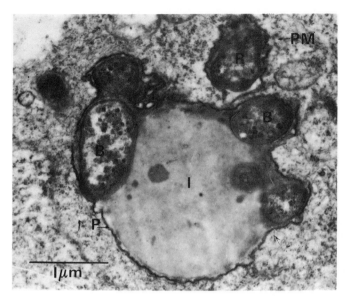

Figure 2.5 Electron micrograph of bacteria leaving the infection thread within a pea nodule. The plasmalemma which is being pushed out by the bacteria will eventually close in behind them to form the peribacteroid membrane. *I*, infection thread matrix; *B*, bacterium; *P*, plasmalemma; *R*, released bacterium; *PM*, peribacteroid membrane.

larger. However, in other plants such as clover and pea, the bacteroids have typical shapes. In pea, the bacteroids often have a Y-shape (Figure 2.6) while clovers contain pear-shaped bacteroids. They are much larger than the bacteria—pea bacteroids show about a 40-fold increase in volume. The host's organelles are pushed to the periphery of the cell and become concentrated at the corners of the cell next to the intercellular spaces.

As well as a change in bacterial morphology to form bacteroids, there is also a difference in biochemical constitution. The bacteroids change their cytochrome components to cope with the low oxygen concentrations within the nodule; also, of course, the synthesis of the nitrogen-fixing system takes place (Appleby, 1969). The most noticeable change in the host cell at this time is the synthesis of leghaemoglobin. The function and synthesis of leghaemoglobin are discussed in Chapter 4.

Some of the cells in the centre of the nodule remain uninfected, and in some nodules which synthesize ureides for nitrogen transport, the synthesis of these compounds takes place in these uninfected cells (see Chapter 6).

The root nodule is well adapted for nitrogen fixation in air. It has a cortex

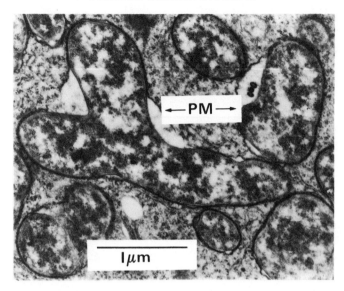

Figure 2.6 A bacteroid from a pea nodule. Note that the peribacteroid membrane, *PM*, is closely appressed over most of the surface of the bacteroid.

surrounding the central core of infected cells, and this contains a layer of cells which act as a diffusion barrier to hinder the entrance of excess oxygen and thus maintain an environment low in oxygen in which nitrogen fixation can take place. The cortex is well supplied with vascular tissue necessary to convey the substrates for nitrogen fixation, nodule maintenance and growth and also to convey the products of fixation (nitrogen-containing compounds) to the shoot and root system for the growth of the plant. The vascular system has to cope with a large flux of compounds moving in two directions and this is assisted by transfer cells in the phloem.

2.1.7 *Effective nodules*

Effective nodules can be defined as those nodules which fix sufficient nitrogen to be of benefit to the host plant. Not all host–*Rhizobium* strain combinations are equally effective—some fix more nitrogen than others (Table 2.1). The reasons for this are complex and effectiveness may depend upon environmental factors. For example, one strain may be most effective if the roots are maintained at a low temperature; at higher temperatures another strain may be the more effective. Effectiveness may also have a more straightforward basis; for example, strains that possess the uptake

Table 2.1 Different degrees of effectiveness of a number of strains of *Rhizobium trifolii* on subterranean clover grown in soil at pH 5.9. The total nitrogen is that which is accumulated in the plant after 11 weeks' growth. Data taken from Thornton and Davey (1983).

Strain	N8	N5	S6	S36	B14	LS5	C100
Total N (mg/plant)	3.7	3.4	3.0	2.8	2.5	2.2	2.0

hydrogenase may be more effective than similar strains which lack this enzyme (see Chapters 3 and 4).

2.1.8 Ineffective nodules

The formation of the nodule can be seen to be a complex process. It may break down at any stage, and if it does so an ineffective nodule is formed. Ineffective nodules either do not fix nitrogen, or else fix so little that the plant does not derive any benefit. Although a great nuisance to agriculture, strains which form ineffective nodules are of value to workers who seek to understand the underlying processes of nodule formation. There is a wide range of defects in ineffective nodules, varying from the inability of the bacteria to emerge from the infection thread to the production of nodule tissue with bacteroids of nitrogen-fixing ability in which the fixing cells senesce prematurely so that they do not last long enough in an active state to be productive.

2.2 Actinorhizal root nodules

Species that associate with the actinomycete *Frankia* have been identified in 24 genera in eight angiosperm families (Figure 2.7). These eight families are not closely related, but they are all dicotyledons and the species that are nodulated are all shrubs or trees (Bond, 1983). The economic importance of host–*Frankia* associations does not rival that of the legume–*Rhizobium* association, although the use of actinorhizal plants is becoming increasingly important. *Casuarina* is widely used as a source of tropical firewood and of poles for constructional use in underdeveloped countries, and is employed in tropical and subtropical regions for sand and soil stabilization and reclamation. In temperate regions, some alder species are of regional importance for timber production, for example *Alnus rubra* in the northwestern United States. Several actinorhizal nodulated species have been employed to provide nitrogen input into forestry ecosystems whereby some of the nitrogen that they fix becomes available to trees that are grown with

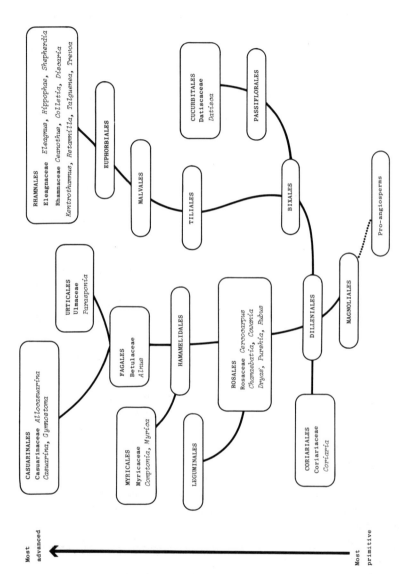

Figure 2.7 Relationships within the Lignosae (woody dicotyledonous angiosperms) between the known orders of actinorhizal plants, based on the phylogenetic system of Hutchinson (1973). Orders are shown in upper case, families in lower case and genera in italics.

them. They have often been used experimentally in short-rotation 'energy' plantations, because of their ability to fix nitrogen. Many species of actinorhizal plants are pioneers and colonize denuded sites that are nitrogen-poor. This ability has led to the use of some of these associations for amenity purposes to restore derelict land (Dommergues *et al.*, 1984; Gordon, 1984).

There are a number of differences between actinorhizal root nodules and those of the legume, and also some similarities. A comparison between the two types can therefore be instructive and from this one can learn something about symbiosis and nitrogen fixation.

Because work on *Frankia* and its hosts has been less active than on the *Rhizobium* symbiosis, only a few host–endophyte combinations have been studied in detail. Although only a few species of host have been examined with regard to the mode of infection and nodule development, some generalizations may be made.

Although the host range of *Frankia* is wide, it is limited to eight families within which there is specificity for infection of different genera, and therefore the problem of recognition has to be overcome. It is probable that this recognition involves lectin binding as has been suggested for *Rhizobium*–legume recognition (section 2.1.1 above). The binding of *Frankia* to the host cell walls has been observed, but the host lectins that may be involved have not as yet been isolated (Chaboud and Lalonde, 1983).

As with the *Rhizobium*–legume association, infection of actinorhizal plants by *Frankia* can be through the root hairs or via the intercellular spaces of the epidermis. The latter mode of infection has been shown to occur in the Elaeagnaceae by Miller and Baker (1985). Hyphae penetrate between the epidermal cell walls of the roots and grow in the intercellular spaces of the cortex. In this early stage, the infection process bears some resemblance to that in *Parasponia*, where *Rhizobium* enters through the epidermis, in this case at the base of clumps of multicellular root hairs. The infection remains intercellular and infection threads are not formed until the host cells are invaded. In *Elaeagnus*, a nodule primordium is initiated from the pericycle of the root in advance of the colonizing *Frankia*. The centre of the developing nodule is invaded by *Frankia*, which after some further growth through the intercellular spaces penetrates a cell wall. The penetrating hypha is encapsulated in cell-wall-like material and can undergo further differentiation to produce vesicles in which nitrogen fixation occurs.

The first stage of the root hair infection process involves deformation of the root hairs (Figure 2.8). The mechanism by which this occurs has not

Figure 2.8 Root hair deformation in *Alnus* following inoculation of seedlings with *Frankia*. Photograph S. Allison.

been studied in actinorhizal plants, but an interesting co-operation of a third organism in this process has been described. Some pseudomonads have the capability also to deform root hairs, although this is not specific to the hosts of *Frankia*. In *Alnus* roots inoculated with *Frankia* together with *Pseudomonas cepacia*, the number of deformed root hairs is considerably increased compared with the numbers formed following inoculation with *Frankia* alone (Knowlton et al., 1980). The increase in deformed root hairs then increases the number of sites suitable for infection and subsequent nodule formation. The fluorescence of deformed root hairs, when stained with acridine orange, is different from that of normal root hairs. This indicates that the cell wall composition is altered; changes in wall characteristics may be necessary to allow root hair deformation and may also assist in the infection process. *Frankia* enters root hairs at the same type of site as *Rhizobium* and the hyphae entering the root hair are also separated from the cytoplasm by cell wall, but in *Frankia* infections the cell wall adheres closely to the hyphae (Lalonde and Knowles, 1975).

After entering the deformed root hair cell, the hyphae may branch, and they then grow down into the cortex. Host cells there are stimulated to divide and also to enlarge, but they remain diploid. This cell division causes a swelling in the root which, because it is not the nodule proper, is called the prenodule. Cells of the prenodule are invaded by the hyphae. At the same time as the formation of the prenodule, nodule primordia develop in the pericycle opposite the xylem poles. Cell divisions in the pericycle and

Figure 2.9 A nodule cluster, several years of age, from a mature tree of *Alnus glutinosa*.

endodermis initiate the nodule primordia but soon afterwards the cortical cells also divide. The number of primordia formed and whether they branch or not depends upon the species of the host. Since they form in the pericycle, the development is characteristic of lateral root formation, and the actinorhizal nodules resemble modified roots structurally, although they arise between the normal points of origin of lateral roots proper. As the primordia develop they are invaded by hyphae which then continue growing in the cortex while the nodule grows from the meristem at its tip. Meristematic cells and cells of the central vascular cylinder are, however, not invaded (Angulo *et al.*, 1976; Callaham and Torrey, 1977).

The meristem at the apex divides causing dichotomous or trichotomous branching which produces the coralloid form of the nodule. Two morphological types of nodule can be recognized. In the *Alnus* type, the nodule meristem is arrested so that a globose mass of nodule lobes is slowly formed (Figure 2.9). This is the most common morphological form of nodule. In the Myricaceae and Casuarinaceae, however, after a period of arrest the nodule lobe meristem begins new development to form the nodule root (Figure 2.10), a negatively geotropic structure which is always free of

Figure 2.10 Nodules of *Myrica gale*, grown in water culture, showing negatively geotropic roots.

endophyte and which is of importance in oxygen supply to the nodules under conditions of poor soil aeration (Torrey and Callaham, 1978). Within the nodule the infected cells are confined to the cortex, the actual distribution differing somewhat between different plant genera. In nodules such as those of *Alnus*, in longitudinal section (Figure 2.11) the cortical host cells nearest to the meristem can be seen to contain only the hyphal form of the endophyte, but immediately below this zone the infected cells show considerable hypertrophy and spherical swellings or vesicles (Figure 2.12) may be seen at the tips of the hyphae. These are arranged around the periphery of the cell, with the hyphae in the centre. The vesicles have laminated cell walls and, as with the hyphae, remain separated from the host cell cytoplasm by a polysaccharide layer and plasmalemma. As in free-living *Frankia*, nitrogenase is located in these structures. A third structural form — sporangia — may be seen frequently in the more distal regions of the nodule. The sporangia are also surrounded by a polysaccharide layer but this covering does not persist when the spores are released as the cell

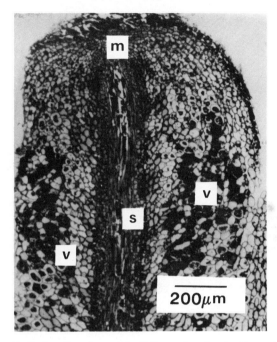

Figure 2.11 Longitudinal section of a nodule of *Alnus rubra* showing the central stele, *s*, the meristem, *m*, and infected cells, *v*, which contain the vesicle form of *Frankia*.

senesces. They are rather angular isodiametric cells which have been called 'bacteroids' and also 'granules' in past literature. They are not to be compared with *Rhizobium* bacteroids as they do not fix nitrogen, and are now believed to be a resting stage of the endophyte. In deciduous plants species such as *Alnus* or *Myrica*, sporangia are more prominent in late summer and winter when nodules lie dormant and the vesicular form of the endophyte degenerates.

Although most *Frankia* strains produce sporangia in submerged culture under the appropriate conditions, not all strains produce spores *in vivo*. The reasons for this are not understood but have led to the typing of some *Frankia* strains as spore (+) or spore (−). The ability to form sporangia in symbiosis is controlled by the endophyte rather than the host plant genotype, although the host may influence the number and seasonality of sporangia formation (Van Dijk, 1978; Vanden Bosch and Torrey, 1985).

Some interesting differences have been observed between the nodule structures of different actinorhizal species. While the infected cells are always

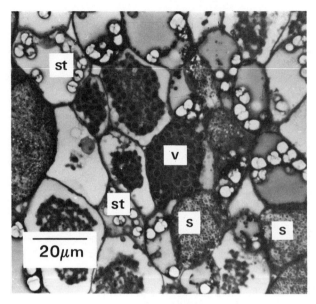

Figure 2.12 Section of the cortex of *Alnus glutinosa* nodule showing infected cells which contain vesicles, *v*, and spores, *s*, of *Frankia*, and uninfected cells which contain starch, *st*.

located in the cortex (Figure 2.13), in some species these cells are confined to a much narrower band in the central (*Myrica*) or outer (*Rubus*) cortex than in *Alnus*. In nodules of *Coriaria* and *Datisca*, the nodules have an acentric stele and the infected cells are asymmetrically distributed in a crescent shape in the cortex (Figure 2.13b). In these genera the vesicles of the endophyte are directed toward the centre of the infected cell instead of the more usual peripheral distribution. Spherical, septate vesicles are typical of the endophyte in nodules of some genera such as *Alnus, Hippophae, Colletia, Dryas, Rubus* and *Discaria*. Elliptical, aseptate vesicles are present in *Ceanothus* nodules. Club-shaped septate vesicles are to be found in *Myrica* and *Comptonia* nodules, while in *Casuarina*, vesicles are not readily recognized although septate, club-shaped tips to some hyphae have been found. All *Frankia* strains form vesicles when fixing nitrogen in culture; however, vesicle form and structure *in vivo* is not a suitable primary taxonomic characteristic for *Frankia* since the host plant can influence its morphology. For example, *Frankia* CpII forms club-shaped sporangia in *Comptonia* but spherical vesicles in *Alnus* (Lalonde, 1979).

The properties of the two symbioses, legume and actinorhizal root

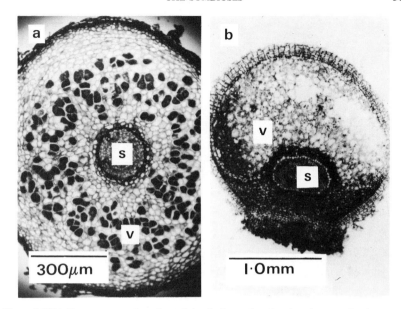

Figure 2.13(*a*) Transverse section of a nodule of *Alnus rubra* showing the central stele, *s*, and infected cells, *v*, which contain the vesicle form of *Frankia* in the cortex. (*b*) Transverse section of a nodule of *Coriaria myrtifolia* showing the acentric stele, *s*, and kidney-shaped area of vacuolate infected cell in the cortex. Kindly supplied by Professor G. Bond and Dr Salma Mian.

nodules, are summarized in Table 2.2. It can be seen that the similar demands which are made upon the different hosts and microsymbionts by the process of nitrogen fixation reduce the scope for differences between these associations. It is clear from the table that the infection process is similar in both types of symbioses. Mutualistic symbiosis has been described as controlled parasitism, and the surrounding of the microorganism by the host cell wall and plasmalemma is probably a defence mechanism to prevent the microorganism from damaging the host's cell contents. The plasmalemma is permeable to small molecules, so that nutrient interchange can take place between the partners, but is impermeable to large proteins, so that enzymes secreted by the endophyte cannot penetrate into the host cell. The first five points on the table are similar, with the assumption that it will be found that lectins are concerned with non-legume–*Frankia* interactions, because they are concerned with the same thing; the formation of bacterial–plant symbiosis. The seventh and eighth points of the table are concerned with nitrogen fixation. In each case the

Table 2.2 A comparison of some aspects of legume root nodules and actinorhizal root nodules.

	Nodule infection and development	Legume nodules	Actinorhizal nodules
1.	Mutual recognition by lectins	+	?
2.	Organism bound to root hair wall	+	+
3.	Deformation of root hairs	+	+
4.	Infection through root hairs	+	+
5.	Organism separated from the plant cell contents by plant cell wall and plasmalemma	+	+
6.	Release of organism from enclosure by cell wall	+	−
7.	Development of specific nitrogen-fixing form	Bacteroids	Vesicles
8.	Haemoglobin present in infected tissue	+	+
9.	Origin of the nodule	Cortex	Pericycle
10.	Increase in ploidy of host cells	+	−
11.	Only polyploid cells infected	+	−
12.	Vascular tissue as strands in the cortex	+	−
13.	Vascular tissue as central cylinder	−	+
14.	Diffusion barrier 'endodermis' present in the cortex	+	−

metabolism is changed so that the process of nitrogen fixation can take place and this change in metabolism is reflected in a change in form of the microsymbiont. The vesicles can be more readily understood as modifications in structure necessary for oxygen protection.

The diffusion barrier present in the cortex of legume nodules is not necessary for actinorhizal nodules where the vesicles provide protection from oxygen. The reason for the variable amounts of haemoglobin in the nodules of different actinorhizal plant genera is not understood, but it is of interest that the largest quantities of haemoglobin are to be found in *Casuarina* nodules, where the typical vesicle structure is either absent or rarely present.

2.3 The *Rhizobium–Parasponia* association

The genus *Rhizobium* was defined as comprising those bacteria which form root nodules with members of the Leguminoseae, and this remained so until Trinick (1973) showed that a tropical tree, now known to be *Parasponia* in the Ulmaceae, was also nodulated by rhizobia. This association is interesting, as the nodules fix nitrogen and share some features with legume nodules and with actinorhizal nodules.

The first stage is the modification of the root hairs, as it is in both the legume–*Rhizobium* association and in the actinorhizal associations. The root hairs in this case are modified somewhat differently. Instead of the root hairs being composed of a single cell they become multicellular. Below the modified hairs, cell divisions occur in the outer layers of the cortex. It is thought that this production of multicellular hairs and cell division serves to produce a stress on the epidermal hairs, causing them to part and bacteria to gain entry into the intercellular spaces through the small gaps that appear. This entry into intercellular spaces is similar to the mode of entry of *Rhizobium* into the roots of *Arachis hypogea* where the penetration is into the intercellular spaces of wounds caused by the emergence of lateral roots. Once within the intercellular spaces, the bacteria divide and cause a loosening of adhesion of the cell walls. Cell loosening and cell division then provide more access points for the bacteria into the root.

Infection of the cells occurs by a method analogous to that in legumes: an infection thread is produced as the bacteria penetrate through the host cell wall. Thereafter, infection of the host cells is by continued infection thread formation.

The structure of the mature nodule is akin to that of actinorhizal nodules, as it is a modified lateral root. Owing to a number of divisions of the apical meristem the modified roots become much branched and coralloid in form.

The bacteria infect cells in the mid-cortex in a similar position to the infected cells in an *Alnus* nodule. The bacteria remain enclosed within cell-wall polysaccharide material. As the nodule matures the character of the polysaccharide changes, as is indicated by a different appearance when viewed with an electron microscope, and at this stage the cells begin to fix nitrogen. It has been suggested that the polysaccharide layer which still encloses the bacteria should be termed an infection thread at this stage rather than a fixation thread. The *Parasponia* nodules are similar to the actinorhizal nodules where *Frankia* remains coated with polysaccharide wall material. It seems possible that this polysaccharide material is concerned not only with the separation of the endophyte from the cell cytoplasm but also with resistance to gaseous diffusion, and the change in structure from the infection thread wall to the fixation thread wall may be required to achieve this.

Parasponia nodules contain haemoglobin. They do not have the pink colour of legume nodules, and it has been suggested that the haemoglobin is confined within the fixation threads (Lancelle and Torrey, 1984; Price *et al.*, 1984).

2.4 Symbioses with cyanobacteria

The range of plants with which cyanobacteria can form a symbiosis is very wide indeed. It encompasses diatoms, fungi, mosses, liverworts, ferns, cycads and angiosperms. The cyanobacterial symbiont has not always been identified with certainty, but is usually a species of *Anabaena* or *Nostoc*, both of which are heterocystous forms.

More often than not the proportion of heterocysts to vegetative cells is much higher in the symbiotic form than in the free-living cyanobacteria. Studies on the differentiation of heterocysts in the filaments has shown that this is determined by the nitrogen status. Nitrogen is fixed by the heterocysts and diffuses into the vegetative cells. As these cells divide the distance of the distal cell from the heterocyst increases and the available nitrogen is depleted by the demands for growth. When this reaches a low level a new heterocyst is developed. The nitrogen compound that is 'sensed' to indicate the nitrogen status is thought to be glutamine, and the concentration of this compound controls the development of the cell into a heterocyst. In the symbiosis where the products of nitrogen fixation are excreted for use by the host, combined nitrogen levels along the filament become depleted sooner than in free-living forms, thus resulting in a greater proportion of heterocysts in symbiotic systems. As with other symbioses, there are morphological adaptations of the host and also physiological adaptations to cater for the special demands of nitrogen fixation.

2.4.1 *Lichens*

Lichens are symbioses of fungi, ascomycetes and basidiomycetes with algae. By far the highest proportion of lichen species are associations of fungi with green algae, but about 25 genera have cyanobacteria as the 'algal' symbiont. There are also a few species, such as *Peltigera aphthosa* and *Stereocaulon vesuvianum*, that have a triple association. The main part of the thallus is composed of an association of the fungus with a green alga and special structures, cephalodia, contain cyanobacteria which fix nitrogen for the thallus as a whole.

The structure of the lichen thallus varies from *Collema*, where the fungal hyphae and cyanobacterial filaments intermingle through the thallus, to *Peltigera canina*, where the cyanobacteria are confined to a thin layer beneath the upper cortex of fungal hyphae. In the lichens the form is determined by both the fungus and the alga, so that it is not possible to ascertain whether any particular change in form is concerned with the cyanobacterial association or to nitrogen fixation.

Lichens that contain cyanobacteria are of ecological rather than of economic importance. The possession of nitrogen-fixing ability by some lichens must be assumed to be an advantage; however, the cyanobacterial associations with fungi do not seem to have any great advantage over the green algal associations except in conditions of extreme nitrogen deficiency. The nitrogen input to the habitat in which they live is also in general small, although it might be significant in the tundra where lichens are dominant and growth of all organisms is rather slow.

The fungus depends upon the cyanobacterium for both carbon and nitrogen when the sole partner is a cyanobacterium, but for nitrogen alone when the cyanobacteria are confined to cephalodia. When the cyanobacteria are present in the main thallus the proportion of heterocysts is the same as in the free-living cyanobacteria. This is because the growth of the lichen is slow and nitrogen is not the growth-limiting factor. When the cyanobacteria are confined to the cephalodia, the nitrogen is exported to the main thallus. This forms a large sink for the nitrogen, as it is required both for the fungus and its green algal partner. As the cephalodia form a small proportion of the whole organism, the nitrogen is depleted and as a consequence the proportion of heterocysts is high.

In order that the fungus may benefit from the nitrogen-fixing ability of the cyanobacteria, the nitrogen metabolism of the latter becomes modified (Stewart and Rowell, 1983). The enzyme concerned with the incorporation of nitrogen in free-living cyanobacteria is glutamine synthetase. The amount of this enzyme in the endophytic form of the cyanobacterium is considerably reduced so that it is unable to incorporate much of the nitrogen that it has fixed and ammonia is exported to the fungus. Amounts of other nitrogen-metabolizing enzymes concerned with the further processing of nitrogen compounds for amino-acid synthesis are also reduced in the endophyte. The mechanism by which this reduction in nitrogen-metabolizing enzymes is achieved is not as yet known.

The evidence for the role of lectins in legume–*Rhizobium* recognition has led workers to look at lichens for similar recognition proteins. Lectins from the fungus have been found which bind specifically to the cyanobacterial partner.

2.4.2 *Azolla*

Azolla (Figure 2.14) is a small fern which grows on the surface of water. It has a prostrate branching stem on which the deeply bilobed leaves are alternately arranged. The lower leaf lobe floats on the surface of the water and thus enables the upper lobe to remain in an aerial position above the water. The adventitious roots, through which mineral uptake occurs, hang down into the water. The plants are usually found in very slow-moving or

Figure 2.14 A plant of *Azolla* sp.

still water as they are fairly fragile, and although they can regrow if fragmented, overall growth is slowed down. *Azolla* is a tropical genus and thus cannot withstand frosts. In hot weather it is prone to insect attack and disease. However, the sporophyte has to be kept in cultivation during the hot summer in order to produce the following crop as methods of inducing sporulation, harvesting and germinating the spores have yet to be developed. This makes the management of *Azolla* difficult and time-consuming. In spite of this, it is used extensively in the Far East for the provision of nitrogen for rice crops.

Azolla is applied to rice in two ways. It can be grown separately and ploughed in as a green manure, or it can be applied to the paddy fields at planting time and grown until it is shaded out by the rice. It can also be applied in both ways to a rice crop. Increases in yield of rice of about 20% can be obtained when *Azolla* is applied. Research in China is currently investigating the best species to suit particular agricultural conditions and also selecting the best strains of cyanobacteria to incorporate into the symbiosis to obtain the most efficient nitrogen fixation. The plant grows vigorously and under good conditions has a doubling time of about two days. A good crop of *Azolla* can accumulate as much as 10 kg N/ha/day but the amount of nitrogen that can be fixed is limited, since the growing season is restricted to the cooler winter or spring due to the disease and insect problems brought by the hot weather. Because of this, it is managed so that it will be shaded out by the growing rice crop in early summer. Part of the time will be spent in the development of *Azolla*, so that although fixation rates are high the amount fixed per season is not as high as 10 kg N/ha/day would suggest. However, its usefulness is indicated by the fact that *Azolla* has extended to cover over 1.5 million hectares in China and Vietnam.

The endophyte of *Azolla* is *Anabaena azollae*. Filaments of *Anabaena* which have no heterocyst and are thus not nitrogen-fixing are associated with the apical meristems of the main rhizome and branches of *Azolla*. They are enclosed in the lobes of a differentiated leaf which protects both the *Anabaena* filaments and the meristem from the environment. In the leaves which develop from the meristem, a small depression forms in the basal surface of the upper leaf lobe which has associated with it a branched hair that grows towards the apical *Anabaena* filaments. The hair has a transfer structure with deeply infolded walls which increase the surface area of the cell membrane. This suggests that it is concerned with the transfer of metabolites to and from the endophyte. The *Anabaena* cells remain near the branched hair and the epidermal cells of the leaf enlarge to enclose the cavity caused by the depression in the leaf surface. Thus both branched hair and *Anabaena* filaments become enclosed within the upper leaf lobe. At this time a further branched hair develops in the cavity and then a number of simple hairs. These all have transfer cell structure. The branched hairs are adjacent to the leaf traces and the simple hairs more randomly arranged, and on the basis of their distribution it has been suggested that the simple hairs are concerned with the transfer of sugars to the endophyte while the branched hairs are more concerned with the absorption of ammonia released by the endophyte into the cavity. As the epidermis closes over the cavity, heterocysts develop on the *Anabaena* filaments and nitrogen fixation commences (Peters *et al.*, 1984).

Modifications in the *Azolla* host plant involve the production of a leaf with a cavity and modified cells, the hairs, which facilitates the interchange of molecules between the partners (Figure 6.9). *Anabaena* does not undergo any morphological changes other than to produce an increased proportion of heterocysts.

However, in this symbiosis biochemical changes are evident which enable the host to benefit from the symbiosis. Nitrogen fixed by the endophyte is released into the cavity as ammonia where it is taken up and transferred to the growing apex of the host, and also to the growing

Anabaena at the stem tip which also depends upon this nitrogen for growth. As in the lichens the release of the newly fixed nitrogen is caused by the fact that the endophyte cannot use most of it for itself. The normal route of nitrogen assimilation in *Anabaena* is by glutamine synthetase. The endophyte has very little, if any, of this enzyme, and over 90% of glutamine synthetase activity in the symbiotic association is found in the host. Thus the ammonia that the *Anabaena* cannot use will leak into the cavity where it is available for the host. The *Anabaena* filaments in the leaf activity have a high frequency of heterocysts which lack photosystem II and thus cannot fix carbon. Sucrose must therefore be obtained from the host, and this can be metabolized to supply reductant for the high nitrogen fixation rates (Peters *et al.*, 1984).

2.4.3 Cycadaceae

The cycads have long been used as ornamental plants in developed countries but are of limited economic importance. Some species indeed are eradicated, as their leaves are poisonous to cattle. However, a role for cycads has been clearly established in the *Eucalyptus* forests of Western Australia, where management involves periodic burning of the understorey to regulate litter accumulation and thus reduce the risk of woodfires. Over 70% of the nitrogen in the litter may be lost by burning, three-quarters of which may be replaced in the 5–7-year period between burns by nitrogen fixed by the cycad *Macrozamia* (Grove *et al.*, 1980). Colonization of the eucalypt understorey by *Macrozamia* is illustrated in Figure 2.15. The cycad association has a number of points of special interest. The symbiosis does not seem to be very specific with regard to the cyanobacterial partner as even with one species of cycad, *Macrozamia communis*, both *Nostoc* and *Anabaena* have been isolated from the root nodules.

The root nodules, like those of the actinorhizal symbiosis, are composed of modified roots, and the gross morphology of the nodules is rather similar. The modified roots become apogeotropic and grow up towards the surface of the soil, branching dichotomously or trichotomously to form coralloid clusters. Cross-sections of the root nodules show that the cortex is in two parts with a distinct inner and outer cortex. The endophyte is confined to the intercellular spaces between the cortical layers. Anatomical investigation has shown that the outer cortex is in fact a persistent root cap. The root cap does not slough off as in normal roots, but remains and grows with the root. The inner layer of the root cap cells does not keep pace in elongation with the cells of the inner cortex, so that intercellular spaces are

Figure 2.15 Colonization of *Eucalyptus* understorey in Western Australia by *Macrozamia riedlii*. Photograph kindly supplied by Dr N. Malajczuk.

formed which are later infiltrated with the endophyte, and the endophyte is thus confined to one layer within the root (see Figure 6.10*a*, *b*).

The infection process has proved difficult to investigate. It appears that cyanobacteria penetrate the cortex at a break in the epidermis of a young root before the periderm has laid down suberized cells. Whether these are accidental breaks, caused by unequal plant growth, or are caused by the cyanobacteria, has not been determined. Having penetrated into the cortex, the cyanobacteria migrate through the intercellular spaces or may digest their way through the middle lamella to reach the spaces between the root cap, outer cortex, and the root inner cortex.

In the lichens, *Azolla* and most cycad systems, the cyanobacteria exist outside the host's cells, but in one species of cycad, *Macrozamia communis*, the cyanobacteria have been found inside the cells (Nathanilez and Staff, 1975). This is similar to the symbiosis with *Gunnera* to be considered next. The ultrastructure of these infected cells has not yet been examined, so that we do not known whether the endophyte remains separated from the cell contents by wall material as in the *Frankia* symbiosis, whether it is surrounded by membrane as in the *Gunnera* symbiosis, or is without either.

Whether the cyanobacterial cells are inter- or intracellular in the cycad symbiosis, they still retain pigment, even when the nodules are below the soil surface so rendering photosynthesis impossible. The cyanobacteria thus live heterotrophically on carbohydrate supplied by the plant.

2.4.4 *Gunnera*

Gunnera is a genus of about 50 species in the Haloragaceae, the water milfoil family. It is distributed in the southern hemisphere, mainly in the tropical and subtropical regions. The species vary from those of about 6 cm in overall height to those having large palmate leaves over one metre across. The cyanobacterial symbiont is *Nostoc*, and this symbiosis is of interest as it is the only known cyanobacterial association with an angiosperm. The symbiosis is present in all species of *Gunnera* so far examined, and it is probable that all species in this genus are associated with cyanobacteria.

The genus does not contain any plants of economic importance. Some species are grown in gardens for decorative purposes—the flowers are very small but *G. manicata* is a very handsome plant (Figure 2.16). The amount of nitrogen fixed has been estimated on a few occasions and estimates vary from 20 to 76 kg N/ha/year.

The *Nostoc* enters glands found on the stem in the axils of the leaves. There are three such glands associated with each petiole base: one situated centrally on the stem above the emergence of the petiole and the other two lying laterally on either side of the petiole. The *Nostoc* grows into the stem tissue in a fan-like form so that cells of *Nostoc* invade host cells well within the stem tissue. The infection process has not so far been investigated. The filaments invade the host cells and become intracellular. Inside the cells they are separated from the host cell's contents by the plasmalemma. After having invaded the host cell, further growth occurs and a large part of that growth involves formation of heterocysts. The development of heterocysts continues through the life of the plant cell until cell senescence, when about 60% of the *Nostoc* cells are heterocysts. However, at this late stage nitrogen fixation has ceased.

In the earlier stages, when heterocysts form 20–40% of the *Nostoc* cells, the high proportion of nitrogen-fixing cells provides very active nitrogen fixation, and it has been estimated that this is several times the rate of

Figure 2.16 *Gunnera manicata* growing in Edinburgh. The ruler, for scale, is 50 cm.

nitrogen fixation of free-living *Nostoc* in which about 5% of the cells are heterocysts.

The glands that contain *Nostoc* are shaded by the petiole and by other leaf bases, and the endophyte is also covered by a number of host cells, so that there is very little light for photosynthesis. However, because phycocyanin, a pigment that is associated with photosystem II, is absent from both the heterocysts and the vegetative cells, *Nostoc*, in symbiosis, must live heterotrophically on carbon fixed by the host. When *Nostoc* is separated from the host it is found that light can stimulate nitrogen fixation. This shows that photosystem I is operative even though it cannot be used in the symbiotic condition. The mechanism by which nitrogen is transferred to the host has not so far been investigated, but by analogy with the other symbioses that have been considered, one would expect that it to involve the supression of glutamine formation by the endophyte. When isolated from the host the *Nostoc* can be grown, showing that when the host's influence is removed phycocyanin can be synthesized and that it can then also assimilate the nitrogen that it fixes.

The cyanobacterial symbioses have a wide range, and from a consideration of some of the associations it can be seen that there is a great diversity of interactions, ranging from the lichens where the cyanobacterial partner photosynthesizes and provides the host with both carbon and nitrogen, through the association with *Azolla* where photosynthesis takes place but carbon is supplemented by the host, to the situation in cycads and in *Gunnera* where the endophyte is dependent wholly upon the host for carbon (Silvester, 1976).

Nitrogen has to be given up to the host if the symbiosis is to be mutualistic, and the high proportion of heterocysts in all these associations shows that it is. In the two associations where the mechanism of nitrogen transfer from endophyte to host has been tested, it has been shown to be the same.

2.5 Symbiosis and nitrogen fixation

It is clear from the discussion above that our knowledge of some of the symbiotic associations is very scanty, and only when more is known of these will we really be able to draw together the common threads to permit a sounder understanding of the systems.

Symbiosis is the result of interaction of the two partners, and it is clear from the preceding discussions that in most cases both partners are modified in some way to achieve this. This must arise by the interchange of molecules between the partners, which may act as signals which cause the partner to modify itself or which may themselves cause modifications. The complexities that can arise are indicated by the study of the synthesis of leghaemoglobin, to be discussed in Chapter 4.

CHAPTER THREE

BIOCHEMISTRY OF NITROGEN FIXATION

Nitrogen is a very inert substance due to the fact that the two nitrogen atoms are joined by a triple bond which gives the molecule great stability. The molecule thus has a very high dissociation energy of 9.46×10^2 kilojoules per mole, but with an appropriate catalyst it can be reduced to ammonia. This is done industrially with variations of the Haber–Bosch process. The reaction below is exothermic when standard conditions are used.

$$N_2 + 3H_2 = 2NH_3(aq) \quad \Delta G = -53kJ \tag{1}$$

Examination of the equation shows that the reaction will proceed more to the right if pressures are raised. In order to obtain a good yield in a reasonable time a temperature of 400–500 °C is employed with pressures up to 200 atmospheres.

The synthesis of ammonia from its elements is a very energy-intensive process. The synthesis of 1 imperial ton (2000lb) of ammonia consumes about 31 000 cubic feet of natural gas, 5.59 barrels of fuel oil or two tons of coal. Part of this energy is used to furnish the high temperature and pressure which are required to operate the system. The remaining 60% is used to provide the hydrogen necessary for the reaction (Nichols et al., 1980).

The biological process of nitrogen fixation is also very energy-intensive although it takes place at atmospheric pressure and at ambient temperature. In the biological system energy is required to supply hydrogen reductant, and also energy is required for an operating system, ATP, in the nitrogenase reaction. Because the energy required in the reductant is high, the electron donors need to have a very low, negative, reducing potential.

The reaction, equation (1) above, requires reductant and nitrogen, and in a biological system one would not predict that anything else was necessary.

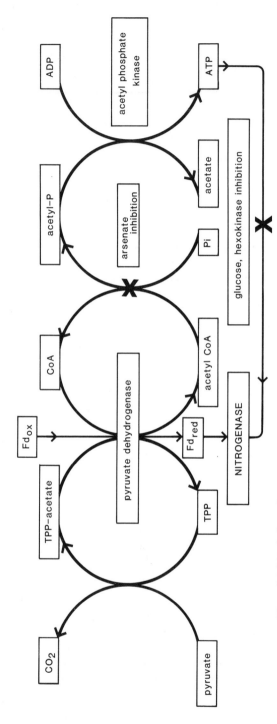

Figure 3.1 Metabolism of pyruvate by the phosphoroclastic system and its interrelationship with nitrogenase. The crosses indicate the points of inhibition by arsenate and glucose + hexokinase.

This held up progress in the subject for a long time, as it was not until the additional requirement for ATP was identified in the early 1960s that an efficient cell-free system could be developed and the nitrogenase enzyme isolated and studied.

The first cell-free extracts, which were obtained from *Clostridium pasteurianum*, could fix nitrogen only when they were supplied with pyruvate or α-oxobutyrate. The pyruvate is metabolized to yield acetate and reductant, ferredoxin. The fact that nitrogen could be reduced only if pyruvate or α-oxobutyrate was added and that other reductants were not utilized, indicated that some other factor which was connected with pyruvate catabolism was also necessary. The nature of this factor was resolved by McNary and Burris (1962) who found that in this system nitrogenase was inhibited with arsenate. Arsenate was known to inhibit the synthesis of acetyl phosphate which in turn is involved in the synthesis of ATP. They also found that, in these crude extracts, added glucose inhibited nitrogen reduction. The explanation for this is that hexokinase, present in the extract, synthesizes glucose-6-phosphate. This uses up any ATP formed by pyruvate metabolism which is thus denied to the nitrogenase.

$$\text{glucose} + \text{ATP} \xrightarrow{\text{hexokinase}} \text{glucose-6-P} + \text{ADP} + \text{Pi} \qquad (2)$$

This work made it fairly certain that ATP is a requirement for the nitrogen-fixing system, a fact that was demonstrated soon afterwards.

The reductant, ferredoxin, is the biological cofactor with the lowest reducing potential E'_0 about -400 mV, which is close to the hydrogen electrode at pH 7.0. This bacterial ferredoxin is related in structure to the ferredoxin found in the photosynthetic system. It is a non-haem iron protein (see below).

3.1 Non-haem iron proteins

As the name suggests, these proteins contain iron but not a haem group. They are concerned with oxidation–reduction reactions and contain an active centre of one or more iron–sulphur clusters (Figure 3.2). The number of iron atoms in a cluster may vary from 2 to 6. More complex clusters may be formed: the iron–molybdenum cofactor, FeMoco, which is present in nitrogenase is a case in point (Figure 3.2*b*).

The iron atoms are attached to the protein by bonding to the sulphur atoms of the cysteine molecules in the polypeptide chain. Each iron atom is also bound to a sulphur atom which bridges two iron atoms to give a cubic lattice. This sulphur is labile and is released as H_2S when acid is applied to the protein.

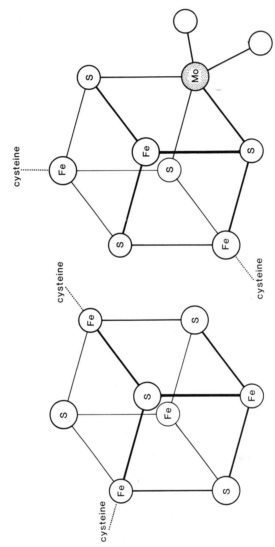

Figure 3.2 The cubic form of iron–sulphur clusters. (Left) FeS cluster. Each pair of iron atoms is bridged by a sulphur atom. Each iron atom is attached to the protein by an S linkage to cysteine (not all shown). (Right) A possible configuration of molybdenum in the corner of an iron–sulphur cluster.

In the oxidation-reduction reactions with which non-haem iron proteins are concerned, oxidation and reduction is achieved by changing the charge on the iron atoms, just as it is in the cytochromes. The energy of the electron held, indicated by the midpoint oxidation–reduction potential E'_0 of the protein, varies widely among different non-haem iron proteins, but as the Fe–S clusters are the same or very similar, E'_0 is controlled by the environment of the cluster within the structure of the protein. Non-haem iron proteins may also contain other prosthetic groups, such as flavin in NADH and succinic dehydrogenases.

3.2 Nitrogenase

The proteins with which we are concerned are ferredoxin and the two nitrogenase proteins, which are all non-haem iron proteins. The nitrogenase proteins were originally purified on an ion exchange column which separated them into two fractions, neither of which had any activity by itself but was active when they were recombined. They were designated 'component 1' and 'component 2' to denote the order of leaving the column. This designation is still used today and is often prefixed by letters indicating the species from which the protein was isolated. Thus kp2 indicates component 2 derived from *Klebsiella pneumoniae*. As they act in an opposite sequence to their component number this seems to be a confusing nomenclature, and it tells one nothing about the protein. Here they will be called the Fe protein (component 2) and the MoFe protein (component 1) respectively, which reminds one of their active centres. The Fe protein has only iron–sulphur clusters but the MoFe protein has iron–molybdenum–sulphur clusters (Figure 3.2b).

3.2.1 *Fe protein*

The Fe protein is the smaller of the two and has two identical sub-units of MW 30 000–36 000 daltons, with one 4Fe–4S iron–sulphur cluster. As the two sub-units are identical in amino acid composition, the cluster must reside between the two sub-units. The Fe protein is extremely oxygen-sensitive and has a half-life in air of 30–45 seconds. Although there are four Fe atoms which can change their valency ($Fe^{3+} \rightarrow Fe^{2+}$) the Fe protein is capable of single electron transfer only. On this protein there are two binding sites for MgATP which are necessary for the transfer of electrons from reduced Fe protein to the MoFe protein. The Fe proteins of any one species of bacterium can associate and react with the MoFe proteins of some, although not all, other species (Emerich and Burris, 1978). The Fe proteins from different species differ slightly in molecular weight. This shows that critical differences must exist in the structures of the Fe proteins from different species, although they all catalyse the same reaction.

3.2.2 *MoFe protein*

This larger protein has two types of sub-unit, each as a dimer, making four sub-units in all, with molecular weights ranging from 50 700 to 60 000 daltons. There are two Mo atoms per molecule in two Mo–Fe–S clusters and also a variable number of Fe–S clusters depending upon the bacterium from which the protein was isolated. Experimental difficulties associated with the contamination of the protein during purification, and also loss of Fe–S clusters during this process, have meant that the molecular weight of the MoFe protein is not known with certainty and neither is the number of Fe atoms associated with it. The MoFe protein is also very oxygen-sensitive, but less so than the Fe protein: it has a half-life in air of about 10 minutes.

3.2.3 *Molybdenum cofactor*

The MoFe–S cluster can be isolated from nitrogenase. Extracts which contain this were found to activate mutant nitrate reductase proteins which lacked a molybdenum cofactor. This gave rise to the concept of a molybdenum cofactor common to all molybdenum redox proteins. However, this was later found not to be so. The iron molybdenum cofactor, FeMoco, was initially contaminated with another molybdenum compound, molybdenum cofactor, Moco, which could activate other molybdenum-requiring proteins. Both FeMoco, which is specific to the MoFe protein, and Moco, which appears to be the common cofactor of the other molybdenum enzymes, have similar properties. They are both oxygen-sensitive and are also very labile in water. They differ in molecular weight. The finding of Moco together with FeMoco would suggest that perhaps Moco is a precursor of FeMoco in nitrogen-fixing organisms. In spite of the fact that FeMoco has a simple elemental composition, its structure has not been completely defined.

One very interesting fact regarding this cofactor is that it can carry out one of the reductions of nitrogenase (acetylene to ethylene) without the presence of the protein, although not as actively. Sodium borohydride was the reductant.

3.3 The mechanism of nitrogenase

The overall reaction is the acceptance by the Fe protein of electrons from an electron donor and the subsequent reduction of the MoFe protein with hydrolysis of ATP. The reduced MoFe protein then reduces the substrate. There is evidence that the Fe protein has two MgATP molecules bound to it when it is reduced. Magnesium is necessary for binding. The overall reactions can be summarized in the following equations:

$$Fe_{ox}(MgADP)_2 + 2MgATP + SO_2^- \rightarrow Fe_{red}(MgATP)_2 + 2MgADP + HSO_2^- \quad (3)$$

$$Fe_{red}(ATP)_2 + MoFe_{ox} \rightarrow Fe_{red}(MgATP)_2 MoFe_{ox} \quad (4)$$

$$Fe_{red}(ATP)_2 MoFe_{ox} \rightarrow Fe_{ox}(MgADP + Pi)_2 MoFe_{red} \quad (5)$$

$$Fe_{ox}(MgADP + Pi)_2 MoFe_{red} \rightarrow Fe_{ox}(MgADP)_2 + 2Pi + MoFe_{red} \quad (6)$$

$$MoFe_{red} + substrate \rightarrow MoFe_{ox} + substrate_{red} \quad (7)$$

The Fe protein binds ATP and is reduced with the liberation of ADP (equation 3). The reduced protein then associates with the MoFe protein (equation 4). There are two binding sites for the Fe protein on the MoFe protein and thus two Fe proteins may be able to associate with the MoFe protein at one time. It would be one if the Fe protein is in short supply relative to the MoFe protein. When the complex is formed the electron is transferred from the Fe protein to the MoFe protein and the ATP is hydrolysed (equation 5). The hydrolysis of the ATP will push the equilibrium of the reaction to the right. The nitrogenase complex is then

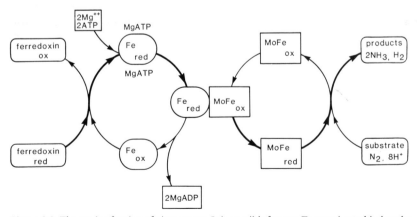

Figure 3.3 The mode of action of nitrogenase. It is possible for two Fe proteins to bind to the MoFe protein at one time; there are two binding sites. Only one is shown for clarity. The pathway for the electron is shown by the thicker line.

dissociated (equation 6) and the electrons are passed on to the substrate (equation 7) and conditions are returned to the start of the series of reactions. Thus a minimum of 2 ATP molecules is hydrolysed for each electron transferred (Figure 3.3) (Low and Thorneley, 1984).

Free Fe protein has an E'_0 of $-294\,\text{mV}$, and on binding to MgATP this is lowered to approximately $-400\,\text{mV}$ (Zumft et al., 1974). The change in midpoint potential denotes an increase in energy of the electron. This is produced by a change in environment of the Fe–S cluster due to a change in protein conformation when the MgATP is bound. Evidence that a change in shape occurs has been found as the properties of the protein change. It becomes more sensitive to oxygen inhibition, attack by sulphydryl reagents such as DTNB and the Fe in the molecule is more susceptible to binding by iron complexing reagents. This shows that the Fe–S clusters become more exposed in the altered protein.

3.3.1 Substrate reduction

Nitrogenase has a wide specificity and will reduce a number of substrates, although it should be borne in mind that under natural conditions it will come into contact with only two of them: N_2 and H^+. Some of the substrates that can be reduced are given in Table 3.1. Apart from nitrogen and protons, the substrate of particular interest is acetylene, reduced to ethylene. The reduction provides the basis for the estimation of nitrogenase activity by gas chromatography and is discussed below in section 3.5.2.

Table 3.1 Substrates reduced by nitrogenase

Name	Formula	Major products
Proton	H^+	H_2
Nitrogen	$N\equiv N$	$NH_3 + H_2$
Nitrous oxide	$N\equiv N^+ - O^-$	$N_2 + H_2O$
Azide	$[N\equiv N^+ - N]^-$	$N_2 + NH_3 + N_2H_4$
Acetylene	$HC\equiv CH$	$CH_2=CH_2$
Allene	$(CH_2=C=CH_2)$ $(CH_3-C\equiv CH_2)*$	$CH_3-CH=CH_2$
Cyanide	$[C\equiv N]^-$	$CH_4 + NH_3$
Alkyl cyanides	$R-C\equiv N$	$RCH_3 + NH_3$
Alkyl isocyanide	$R-N^+\equiv C^-$	$RNH_2 + CH_4$

*Methyl acetylene a probable intermediate

3.3.2 Hydrogen evolution

The reduction of nitrogen by nitrogenase is not as straightforward as the Haber–Bosch process, as protons are also reduced in the course of the reaction. The proportion of electrons donated to protons and to nitrogen is variable, but the minimum ratio is 25% to protons and 75% to nitrogen, which results in one molecule of hydrogen produced for each molecule of nitrogen reduced. Thus the equation of the whole reaction is more properly set out as

$$N_2 + 8e + 8H^+ = 2NH_3 + H_2 \qquad (8)$$

rather than as in equation (1). Taking into account the hydrolysis of ATP this becomes

$$N_2 + 8e + H^+ + 16ATP \rightarrow 2NH_3 + H_2 + 16ADP + 16Pi \qquad (9)$$

The fact that this is a minimum ratio and is stoichiometric suggests that it is an obligatory part of the nitrogenase reaction.

The ratio of hydrogen to ammonia produced can vary and is increased when conditions are not optimal for enzyme reaction and when the enzyme is inhibited. Nitrogenase action continues, provided that it is supplied with ATP and reductant, because there is always substrate present in the form of protons. Thus if the concentration of nitrogen, as substrate, is reduced the rate of enzyme action is not affected much, in fact it speeds up a little, and a greater proportion of hydrogen appears in the reaction products. Hydrogen is not produced when substrates other than nitrogen are being reduced.

Factors which affect the rate of enzyme action and the proportion of hydrogen produced are as follows.

1. The ratio of component proteins Fe:MoFe: this ratio will normally be optimal *in vivo* but can be affected by water stress, and high oxygen levels may affect the Fe protein first, thus affecting the ratio. In root nodules high oxygen levels are only likely to occur under experimental conditions in which oxygen levels over 21% are applied.

2. ATP:ADP ratio: the lower the ratio the greater the amount of hydrogen evolved. This can come about due to energy starvation when the supply of reductant is also likely to be reduced.

3. pH can affect hydrogen production. This might be expected, as protons are involved in the reaction, and has been shown to be the case *in vitro*. However, the large pH changes necessary to produce a marked change in hydrogen output are not likely to occur *in vivo*.

4. Inhibitors: there are a number of inhibitors of nitrogenase which

might affect the ratio, but two are of special interest. Hydrogen, the product of nitrogenase, will inhibit the reduction of nitrogen but not the activity of the enzyme itself, with the result that an increased proportion of hydrogen will be evolved. Hydrogen does not inhibit the reduction of substrates other than nitrogen. Carbon monoxide also inhibits nitrogen reduction without affecting the electron flow through the enzyme, and thus hydrogen evolution is increased.

Of the reactions that can occur naturally, only a shortage of nitrogen as substrate and hydrogen inhibition can effect an increase in the proportion of hydrogen evolved without also lowering the activity of the enzyme. The concern with hydrogen evolution is that the electrons used for the reduction of protons consume ATP and reducing power, and for the most efficient nitrogen fixation hydrogen production needs to be kept to a minimum. In most organisms this seems to be so, but the bacteroids in legume root nodules produce excess hydrogen (Schubert and Evans, 1976): this will be discussed in more detail later.

3.3.3 Electron transport to nitrogenase

The activated Fe protein has an E'_0 of -380 to -490 mV and so will require electron donors with midpoint potentials of around -400 mV in order that a significant proportion of the protein is reduced. In *Clostridium*, which fixes nitrogen anaerobically, there is no difficulty as reduced ferredoxin is supplied as a result of the phosphoroclastic reaction when pyruvate is metabolized (Figure 3.1). Ferredoxins have a range of midpoint potentials from -330 mV to -460 mV, depending upon the type of ferredoxin and the species from which it is isolated; that in *Clostridium* is -420 mV.

Another low-potential electron donor is flavodoxin. Flavodoxins are proteins with a flavin (FMN), group instead of the iron–sulphur cluster. The oxidation–reduction reaction is effected by changing the flavin from a hydroquinone form to a quinone form. In the low potential flavodoxins this will be from the hydroquinone to the semiquinone form (Figure 3.4). These proteins have midpoint potentials in the range of -419 to -495 mV. Whether a ferredoxin or a flavodoxin is the electron donor to nitrogenase depends upon the species of bacterium. *Azotobacter* uses flavodoxin and *Rhizobium japonicum* uses ferredoxin as reductant (Carter *et al.*, 1980). The versatility of bacteria is shown by the ability of *Clostridium*, which normally uses ferredoxin, to utilize flavodoxin when it is short of iron.

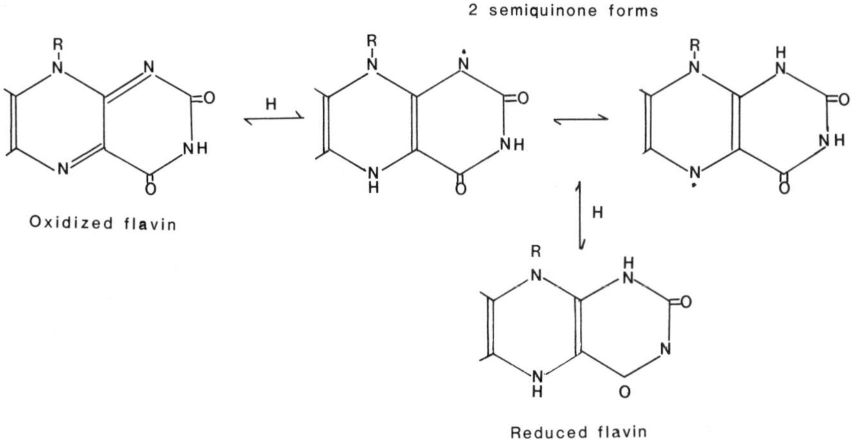

Figure 3.4 The reduction of the quinone form of flavin to the hydroquinone form. The semiquinone has an unpaired electron on the nitrogen atom and is a free radical. It has two resonating forms.

What is of more interest than the actual electron donor to nitrogenase is the mechanism by which the donor is reduced in aerobic organisms. In normal aerobic metabolism, NAD(P)H (E'_0 − 320 mV) is the most reducing compound. In order to maintain ferredoxin at a half-reduced state, at its midpoint potential, the NAD(P)H:NAD(P) ratio would have to exceed 99. This high ratio is not attained in the cell as a number of enzymes, which are controlled by this ratio, such as glucose-6-P, 6-phosphogluconate and isocitrate dehydrogenases, would be inhibited. It is thus not surprising that reduced pyridine nucleotides have been shown to be poor electron donors to nitrogenase in cell-free extracts of *Azotobacter*. Thus *in vivo* there must be some other factor concerned with Fe protein reduction, and the suggestion has been made that extra energy is supplied to electrons by energized membranes in the intact organisms (Haaker and Veeger, 1977). There is a certain amount of experimental evidence in support of this hypothesis, but it is difficult to prove experimentally.

3.4 Hydrogenase

The ATP that is used by nitrogenase will come from normal metabolic processes such as fermentation in anaerobic and oxidative phosphorylation in aerobic organisms. The amount of ATP used is, however, large. Not only is ATP used in the donation of electrons to nitrogen but it is also used in the production of hydrogen from the nitrogenase. Such energy associated with proton reduction is wasted but is necessary if hydrogen evolution is an essential part of the reduction process. Some organisms have an enzyme which can utilize the hydrogen produced by the nitrogenase and are able to regain some of the wasted energy.

Anaerobic organisms possess a hydrogenase which is used to remove excess reducing power. This is a reversible hydrogenase and catalyses the reaction in equation (10).

$$2e + 2H^+ = H_2 \tag{10}$$

Nitrogen fixation in these organisms has to take place with hydrogen present because of the action of this hydrogenase. This hydrogen would be inhibitory but it has been found that the nitrogenase of *Clostridium* is less susceptible to hydrogen inhibition than that isolated from aerobic organisms.

Aerobic organisms may possess an uptake hydrogenase (Robson and Postgate, 1980). This enzyme has been shown to be reversible when purified, but its equilibrium lies very much to one side so that it normally catalyses the reaction shown in equation (11).

$$H_2 \rightarrow 2H^+ + 2e \tag{11}$$

The electrons are passed down an electron transport pathway to oxygen and ATP is regained by oxidative phosphorylation. The net result is

$$H_2 + O + nADP + nPi \rightarrow H_2O + nATP \tag{12}$$

where n is the P/O ratio.

The uptake hydrogenase differs from the reversible enzyme in a number of ways. The reversible enzyme is a non-haem iron protein whereas the uptake hydrogenase contains nickel. The uptake hydrogenase is membrane-bound whereas the reversible enzyme is a soluble component. The aerobic organisms that possess the uptake hydrogenase are the cyanobacteria, *Azotobacter* and some *Frankia* and *Rhizobium* strains.

Nitrogen fixation is often limited by the availability of carbon substrate for energy supply so that one of the roles of uptake hydrogenase is to regain energy lost by hydrogen evolution. Two other possible roles for uptake hydrogenase can be considered. They are:

1. As a substrate for respiratory protection. The utilization of oxygen shown in equation (12) means that carbon substrate that would

otherwise be used for respiratory protection will be spared. The only hydrogen that is available for the uptake hydrogenase will be that evolved by nitrogenase and it has been calculated that this will only contribute to about 6% of the oxygen uptake. The aid that can be given to respiratory protection by this enzyme will then be small.
2. Protection from hydrogen inhibition. Hydrogen produced from the nitrogenase will inhibit the enzyme unless it can diffuse away or is removed. Hydrogen has a very high coefficient of diffusion and can thus diffuse away faster than any of the other gases, so that it has been considered that this role is not necessary. However, this view has been challenged: this is discussed further in Chapter 4.

3.5 Assay of nitrogen fixation

With many metabolic processes the assay of the system is quite straightforward, but the measurement of nitrogen fixation is performed differently depending upon the system that is being analysed and the activity being measured, for reasons that should become apparent as the different methods are discussed. The activity of nitrogenase can be measured directly when an enzyme extract or pure enzyme is used as the product, ammonia, is not metabolized any further. The ammonia is removed from the assay mixture by making the mixture alkaline and absorbing the evolved ammonia in acid. It can then be estimated by titration or colorimetrically.

Three methods that are used on a variety of systems are worth looking at in more detail.

3.5.1 *The Kjeldahl method*

Until the 1940s, when mass spectrometers became available, this was the only way in which nitrogen fixation could be measured. The method involves the digestion of the sample in concentrated sulphuric acid together with a catalyst, thus converting all the organic nitrogen to ammonium sulphate. The digestion mixture is then made alkaline and the ammonia released distilled into standard acid, following which the quantity of acid neutralized by the ammonia can be determined by titration. This method is simple and straightforward but time-consuming, and the sample is destroyed during the analysis. The amount of nitrogen that has been fixed must be large, as it has to exceed the experimental error inherent in the method and the samples. This means that the method cannot be used for short-term experiments but, by the use of appropriate controls, it can be used to measure the amounts of nitrogen fixed over a few days. This method gives accurate results if used with controls and where there is an adequate increment of nitrogen. A number of false claims for nitrogen fixation in mycorrhizas and fungi have appeared in the literature because increases in nitrogen have been recorded when no nitrogen was supplied. The conclusion that these increases were due to the fixation of nitrogen were false because it was not realized that a number of microorganisms are excellent scavengers of nitrogen and are able to accumulate it from the air and from very low concentrations of impurities in the growth medium. No editor now would accept Kjeldahl results alone as sufficient evidence to establish fixation in a new organism. The definitive test for this is the incorporation of ^{15}N from $^{15}N_2$ in the gas phase.

3.5.2 *The isotopic method*

There is a radioactive isotope of nitrogen, ^{13}N, but it has a half-life of only 9.96 minutes so that its application is limited. The other isotope of nitrogen, ^{15}N, is a non-radioactive heavy isotope and must be assayed in a mass spectrometer. The amount of ^{15}N in free nitrogen and normal compounds is quite small, about 0.36%, and as the mass spectrometer is able to measure to an accuracy of at least 0.002%, this is a sensitive and sure way of measuring fixation when $^{15}N_2$ is supplied to the system. It can be used over short time periods if the activity of the nitrogenase is high enough.

To obtain N_2 gas for the mass spectrometer the sample has to be subjected to Kjeldahl digestion and distillation, and the nitrogen is then released by alkaline hypobromite. The isotopic method is sure and accurate but it is very time-consuming and expensive both in terms of the isotope used and the equipment needed.

3.5.3 *The acetylene reduction method*

A considerable advance was made when Dilworth (1966) and Schollhorn and Burris (1966) discovered that acetylene strongly inhibited nitrogenase, and Dilworth showed that the acetylene was reduced to ethylene. These findings introduced a quantitative method that has several advantages over the two methods described above. The method consists of supplying the atmosphere of the nitrogen-fixing system with about 10% acetylene and incubating for the required time. Gas samples are then taken by syringe and the ethylene produced by the reduction of acetylene is assayed by injecting the sample into a gas chromatograph. Each such assay takes less than five minutes.

The advantages of the method are that the substrate, acetylene, is cheap and the instrumentation, the gas chromatograph, is considerably less expensive than a mass spectrometer. Because the assay is very sensitive, short-time incubations can be done; also, because the assay is non-destructive and gas samples can be removed at intervals it is possible to obtain a time course of the reaction on a single sample.

When acetylene is supplied, no protons are reduced, so that the amount of ethylene produced is a measure of the total activity of the enzyme. The reduction of acetylene involves two electrons so that to equate ethylene with nitrogen reduction according to equation (8), four ethylene molecules are the equivalent of the reduction of one nitrogen molecule. Its relative inexpensiveness, ease, speed and non-destructive nature have led to extensive use of this method.

The acetylene reduction method is, however, not without its problems, and these are particularly acute when acetylene is used to measure nitrogen fixation in nodulated systems. The main reason for this is that legume nodules, unless they possess an uptake hydrogenase, evolve a larger proportion of hydrogen than the minimum. Thus unless one knows, or can assay, the proportion of electrons allocated to protons and nitrogen it is not possible to find a factor by which to divide the ethylene value in order to estimate the equivalent amount of nitrogen that would have been fixed.

When soil cores are analysed by the acetylene reduction method there is the additional problem of the time taken for the gases to equilibrate through the pores of the soil. Additionally, some soil microorganisms and plants are capable of ethylene production, which will cause serious errors in the estimation of activity when nitrogen fixation is weak. Soil microorganisms may also oxidize ethylene—a process that is inhibited by acetylene—so that controls set up without acetylene may not give a proper measure of ethylene accumulation when acetylene reduction by soil cores is being measured.

The inability to assess precisely the amount of nitrogen fixed may not matter in comparative tests where the electron allocation is likely to be the same in the experimental objects and the controls. If there is no hydrogenase present, the acetylene assay can be put on a more precise basis by assaying the amount of hydrogen evolved in air. A comparison of this amount with the total electron throughput, measured by acetylene reduction, will give the ratio of electrons donated to protons and to nitrogen. Instead of acetylene reduction the evolution of hydrogen under argon, or other inert gas in place of nitrogen, will measure the total activity of nitrogenase in the system.

Portable gas chromatographs are now available so that nitrogen fixation can be measured in the field. Studies can now be done in natural habitats under ambient conditions to assess the nitrogen input from different nitrogen-fixing systems within the habitat.

Because of the difficulties that are inherent in the acetylene reduction method, the definitive test for nitrogen fixation both qualitatively and quantitatively remains the incorporation and measurement of ^{15}N. The acetylene reduction method has, however, led to important advances in our knowledge and has facilitated nitrogen fixation research considerably.

CHAPTER FOUR

THE BIOCHEMISTRY AND PHYSIOLOGY OF THE LEGUME ROOT NODULE

There are a number of aspects of the biochemistry and the physiology of the developing root nodule that are of interest, but in this chapter the discussion will be confined to those that deal with nitrogen fixation as such rather than with the development of the root nodule as a plant organ. The aspects of physiology to be considered will therefore be concerned with the mechanisms of gaseous exchange and diffusion within the nodule, with the provision of reducing power and ATP to the bacteroids. The transport of solutes to and from the root nodule will be discussed in Chapter 6.

4.1 Gaseous exchange

As the nodule is a respiring system, oxygen and carbon dioxide will be exchanged. In addition the nodule takes in nitrogen and evolves hydrogen, so that the movement of all these gases will have to be considered. Gases diffuse ten thousand times faster through the gaseous phase than through liquid, so that the path length through these two phases will determine the rate at which gaseous exchange occurs. The rate of diffusion of gases into and out of the nodule can thus be controlled by regulating the path length through the gaseous and liquid phases. The layer without intercellular spaces, the endodermis in the cortex mentioned in Chapter 2, prevents the diffusion of excess oxygen into the nodule. This layer was predicted by Sinclair and Goudriaan (1981) who made a mathematical model of the diffusion process. The amount of intercellular space in the nitrogen-fixing tissue will also be important for the control of gaseous diffusion.

The rate of diffusion is governed by Fick's law which states that

$$dw = -D\,dc/dx \cdot dt$$

where w = flux, the quantity diffusing across 1 cm^2 in the time dt with a concentration gradient of dc/dx. As the diffusion coefficient for a gas is a constant at a given temperature, the rate of diffusion will depend upon the concentration gradient. Respiration will rapidly reduce the concentration of oxygen within the nodule to a low level. The result of this will be a large concentration gradient between the inside of the nodule and the outside which will permit rapid replenishment of the respired oxygen. The rate of respiration of the cortex is fairly low, so that the concentration of oxygen in the outer part of the nodule endodermal layer lacking intercellular spaces will differ little from that of atmospheric oxygen. The actual amount will vary with the temperature as the solubility of gases is temperature-dependent, but will be about 240 μM. The concentration of oxygen on the inside of the layer will be very low so that there is a steep concentration gradient assisting diffusion across the liquid layer.

The concentration of nitrogen on the outside is very high, 79%, but as the consumption of nitrogen for fixation is considerably lower than that of oxygen for respiration, the concentration of nitrogen within the nodule is not lowered by a large amount. The gradient for the diffusion of nitrogen will not be as steep as that for oxygen, but as the rate of nitrogen use is smaller it is adequate to maintain nitrogen levels within the nodule so that nitrogenase activity is not limited by nitrogen as a substrate.

The diffusion of substances out of the nodule is a different matter. The concentrations of oxygen and nitrogen outside the nodule are fixed and the rate of diffusion depends upon the depletion of their concentrations inside. The converse is the case for hydrogen and carbon dioxide. In order for these gases to diffuse out rapidly it is necessary to create a concentration gradient sufficient to permit the outward diffusion of these gases as fast as they are produced. Hydrogen has a high diffusion coefficient so that it does not require such a large concentration gradient as the other gases in order to diffuse out at the same rate.

The rate of diffusion of carbon dioxide in aqueous solution is complicated by the equilibrium of carbon dioxide with its hydrated form, the bicarbonate ion. An equilibrium is formed by the following reaction:

$$CO_2 + H_2O = HCO_3^- + H^+$$

The ratio of the two components is given by:

$$\log \frac{[HCO_3^-]}{[CO_2]} = pH - pK \quad (pK = 6.23)$$

At pH 7.0 there is nearly six times as much bicarbonate ion as free dissolved carbon dioxide. The diffusion coefficients for the two components, CO_2 and bicarbonate ion, are 1.94 and 1.019×10^{-5} cm^2 sec^{-1} respectively. Thus, at pH 7.0 and at 15°C, bicarbonate would diffuse three times faster than free dissolved carbon dioxide. However, the hydration reaction is slow so that the rate of diffusion would normally be limited by this.

Root nodules have been shown to contain carbonic anhydrase (Atkins, 1974). This enzyme catalyses the hydration of carbon dioxide and the reverse of this reaction. This means that in nodules diffusion is not limited by the rate of the hydration reaction and is thus governed by the rate of bicarbonate diffusion, which is faster than that of free dissolved CO_2. The catalysis is necessary at both ends of the diffusion pathway, since the diffusion across cell membranes will be faster as free CO_2 than as the charged bicarbonate ion.

The enhanced diffusion as bicarbonate ion means that the free carbon dioxide does not need to attain a high concentration in the centre of the nodule in order to have a concentration gradient sufficiently steep to maintain the same rate of diffusion as oxygen. There is thus less CO_2 to react with leghaemoglobin: the complex of CO_2 with haemoglobin lowers its oxygen affinity. It also means that there will be a high concentration of bicarbonate, which is the substrate for phosphoenol pyruvate carboxylase.

The case for the diffusion of hydrogen within the nodule is interesting, as in order for hydrogen to diffuse out of the nodule it must have a finite concentration within the nodule. As discussed in Chapter 3, hydrogen inhibits nitrogen fixation, with the result that more hydrogen is produced. The question arises as to whether the concentration within the nodule is sufficiently high to inhibit nitrogenase. The solubility of hydrogen in water is about half that of oxygen, but on the other hand the diffusion coefficient is larger; also, the amount of hydrogen produced will be less than the amount of oxygen consumed. If we take an average of four oxygen molecules consumed for each electron pair passed through nitrogenase and one hydrogen produced for every eight electrons passed through nitrogenase, then the rate of hydrogen evolution will be one-sixteenth of that of oxygen uptake. Using the criteria outlined above for oxygen and carbon dioxide, there would be a need for a diffusion gradient equivalent to 2.4% hydrogen in the internal gas phase. This is sufficiently low to be ignored, but this is the concentration gradient necessary across the layer without intercellular spaces. There also has to be a concentration gradient from the centre of the nodule to this point. The steepness of this gradient and thus the highest concentration that will be attained in the centre of the nodule will be determined by the length of the diffusion pathways through liquid and gas in the centre of the nodule.

4.1.1 Intercellular spaces

The proportion of intercellular space in the nodule is critical for its function: such space allows oxygen to diffuse to the centre of the nodule for respiration and to 50 power nitrogen fixation, and permits hydrogen to diffuse away, preventing the build-up of inhibitory concentrations. However, if the proportion of intercellular spaces is critical. If the proportion of intercellular space is too large then the reduced diffusion resistance could allow too much oxygen into the centre. There have been few studies which have attempted to assess the amount of intercellular space in the root nodules. Bergersen and Goodchild (1973) indicated that about 5% of soya bean nodule volume was intercellular space, whereas a study of pea and lupin nodules showed an average of about 1% of the volume as intercellular space. The soya bean root nodule is normally much larger than the nodules of these temperate species, so that in order for the gases to diffuse over greater distances the diffusion paths should be larger. The intercellular spaces in the soya bean are interconnected so that there is a network of space in the interior of the nodule separated from a similar network of space in the cortex. The presence of the network and the interconnection of the spaces has been shown by experiments in which nodules were vacuum-infiltrated with Indian ink. The ink entered the cortex up to the nodule endodermal layer without intercellular spaces and went no further, thus confirming the anatomical studies. When the nodule was cut in half, thus exposing the spaces in the centre to the Indian ink, the ink penetrated through the central zone (Tjepkema and Yocum, 1974). Similar experiments with peas showed that in these nodules the spaces in the cortex were connected, as in soya bean, but that the spaces in the centre of the nodule were not interconnected as the ink did not penetrate into the centre of the nodule. Thus, as the spaces are not connected and the proportion of space is very much less, there is a greater resistance to gaseous diffusion in the pea nodule than in the soya bean nodule.

The effectiveness of the diffusion resistance in preventing oxygen inhibition is shown in experiments in which excess oxygen was applied to root nodules. The graph (Figure 4.1) has been taken from an experiment by F.J. Bergersen, and shows that nitrogen fixation is increased in oxygen concentration up to 50%, after which, at higher concentrations of oxygen, nitrogen fixation declines. Respiration is thus not saturated with oxygen at normal atmospheric pressure, and as the oxygen pressure is increased, respiration is able to increase also, with the result that more nitrogen is fixed. Above a certain external pressure of oxygen, in this

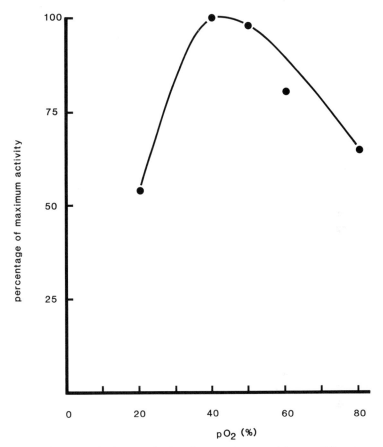

Figure 4.1 The rate of nitrogen fixation in soya bean nodules at different oxygen concentrations.

case about 50%, oxygen can diffuse into the nodule faster than it is consumed by respiration. The concentration will then build up to inhibitory levels and nitrogenase is inactivated. Thus under normal conditions nitrogen fixation in the nodule is limited by diffusion of oxygen. Further evidence for this is that the rate of nitrogen fixation and respiration are not affected by temperature over the range of 15–25°C in pea nodules. Thus over this temperature range they must be controlled by some process relatively independent of temperature, such as diffusion. The diffusion coefficient would not be altered significantly over this range of temperature.

It might seem strange that the nodule has evolved so that it does not fix nitrogen optimally under normal conditions. However, under conditions of low substrate, the rate of respiration will fall and thus the oxygen concentration will rise. There has to be spare capacity if diffusion resistance within the nodule to take care of this circumstance. A similar situation can arise in temperate legumes when respiration rate is lowered by low temperatures; temperatures in soil in temperate latitudes will often be below 15°C during the growing season.

Table 4.1 Changes in nitrogenase activity of pea nodules with temperature

Temperature (°C)	Ethylene production from acetylene
6	10.5 ± 1.2
15	19.9 ± 2.4
25	21.2 ± 2.7

Nitrogenase activity, as acetylene reduction, is not affected by temperature in the range 15°C–25°C. The rate is diffusion-limited.

4.2 Leghaemoglobin

The discussion above has dwelt upon the effects of intercellular spaces on diffusion resistance and the effect this diffusion resistance has on respiration rates and nitrogen fixation. These effects would not easily be attained if it were not for the presence of leghaemoglobin.

Until relatively recently haemoglobins were unknown in plants apart from those in legume root nodules. It is now known that haemoglobins are also present in some actinorhizal nodules and also in the root nodules of *Parasponia*, the non-legume associated with *Rhizobium*. In all these cases the function of the haemoglobin is probably the same.

Leghaemoglobins are monomeric proteins which will reversibly bind oxygen, and are similar in many ways to myoglobin, the red oxygen-carrying protein bound in mammalian muscle. They have one haem group and can bind one oxygen molecule. They have a very high affinity for oxygen and are half saturated at oxygen concentrations of around 10–20 nM. This compares with values of 220 nM for horse myoglobin and 126 nM for the β-chain of human haemoglobin (Wittenberg *et al.*, 1972).

Legume root nodules may contain more than one form of leghaemoglobin; soya bean has four. These forms differ in amino acid sequence and composition but have a number of critical sequences that are conserved and are common to all the leghaemoglobins. These conserved sequences will be associated with the functional parts of the molecule which bind the haem and provide the correct environment to the haem which influences the oxygen binding (Appleby, 1984).

4.2.1 Synthesis of leghaemoglobin

The synthesis of leghaemoglobin is interesting from the point of view of the formation of the functional symbiosis. It involves the cooperation of both partners in the synthesis of a compound peculiar to, and necessary for, the efficient functioning of the association. The polypeptide, globin, is coded for on the plant chromosomes and synthesized on the plant ribosomes. It was demonstrated that this was a plant protein when it was found that different species of legume, *Lupinus luteus* and *Ornithopus sativus*, possessed different leghaemoglobin molecules even though they were inoculated with the same strain of *Rhizobium* (Figure 4.2). A complementary experiment showed that, when one species of plant was inoculated with a number of different rhizobial strains, the leghaemoglobin molecules did not differ with the different inocula.

The synthesis of the haem group was rather more difficult to establish. This is because both the bacteria and the plant require the haem group for other proteins such as cytochromes, so that the specific synthesis of haem for leghaemoglobin must be decided on a quantitative rather than an all-or-none basis. Fortunately leghaemoglobin is at a high concentration, for a protein, within the nodule. In soya beans the concentration throughout the nodule is about 700 μM. It is also synthesized over a short time period in the development of the nodule (Bergersen, 1982). Thus by looking at the

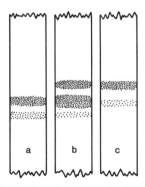

Figure 4.2 Diagrams of the electrophoresis patterns of leghaemoglobins from *Lupinus luteus* and *Ornithopus sativum* inoculated with the same strain of *Rhizobium* a, lupin leghaemoglobin; b, lupin + seradella leghaemoglobin; c, seradella leghaemoglobin. Data from Dilworth (1969).

activities and the sites of the enzymes that synthesize the haem group it has been possible to establish that it is the bacteroids that are responsible for haem synthesis.

Two enzymes that initiate the synthesis of the tetrapyrrole ring, the core of the haem group, are δ-aminolevulinic acid synthase (ALAS):

$$\text{succinyl-CoA} + \text{glycine} \rightarrow \delta\text{-aminolevulinic acid} + CO_2 + H_2O \quad (1)$$

and δ-aminolevulinic acid dehydrase (ALAD):

$$2\ \delta\text{-aminolevulinic acid} \rightarrow \text{porphobilinogen} + H_2O \quad (2)$$

Investigation of the activity of these two enzymes in both bacteroid and plant fractions of the nodule showed that ALAS could be detected only in the bacteroid fraction and that its activity increased as the nodule developed. The second enzyme, ALAD, is found in both plant and bacteroid fractions, but the activity in the plant fraction declines with nodule development while the activity in the bacteroid fraction increases. A further pointer to the fact that it is the bacteroids that synthesize the haem is that ALAD is found only in the bacteroids of nodules where leghaemoglobin is produced and is absent from the bacteroids of ineffective nodules that lack leghaemoglobin. The haem group then must be exported to the plant cytoplasm, where it is united with the globin to make the complete molecule.

The haem synthesis is induced by low oxygen tensions. This has been shown by placing free-living *R. japonicum* cells under low oxygen tension, after which haem synthesis was increased tenfold. The gene for the synthesis of leghaemoglobin is expressed only in the nodule so that it must require some special signal to switch it on. The nature of this is not known.

4.2.2 *Location of leghaemoglobin*

The location of leghaemoglobin within the nodule cell has been a matter of some controversy. If it is to function as an aid to oxygen diffusion (see below) then it should be present throughout the nodule cell. Claims that it is present either only within the peribacteroid membrane or only within the cytoplasm were based on evidence that looked equally convincing. The probability is that leghaemoglobin is distributed much as has been reported by Bergersen and Appleby (1981). Using well controlled extraction methods, these workers showed that the leghaemoglobin is situated both inside the peribacteroid membrane and also within the plant cytoplasm. They were able to centrifuge down intact peribacteroid membrane envelopes and were thus able to find the concentration of leghaemoglobin within them, 340 μM. They calculated the total volume of the cell occupied by the membrane envelopes and by the plant cytoplasm from measurements on electron micrographs. They were thus able to estimate the concentration of leghaemoglobin in the plant cytoplasm as about 3 mM.

Soya beans have several bacteroids within one peribacteroid membrane and there is a large proportion of space between the bacteroids within the membrane. In peas and clovers the peribacteroid membrane is much more closely appressed to the bacteroids and there is little space inside to contain leghaemoglobin.

4.2.3 *Function of leghaemoglobin*

It has been known for a long time that the rate of nitrogen fixation within the nodule is closely correlated with the concentration of leghaemoglobin. This close linkage of leghaemoglobin with nitrogen fixation even led at one time to the suggestion that leghaemoglobin was the nitrogen binding agent.

Leghaemoglobin reversibly binds oxygen and its function will be related to this. There are three possibilities, any of which could be functions of leghaemoglobin:

1. To facilitate oxygen diffusion
2. To act as an oxygen buffer
3. To act as an oxygen donor to specific sites on the bacteroid surface.

The diffusion coefficient of oxygen is about 20 times that of leghaemoglobin (2.35×10^{-5} cm^2 sec^{-1} and 1×10^{-6} cm^2 sec^{-1} respectively) so that at first sight one might not expect that oxygen bound on to leghaemoglobin would diffuse faster than free oxygen. However, the other factor that governs the rate of diffusion is the concentration gradient. The

Table 4.2 Oxygen and leghaemoglobin in root nodules. It can be seen that the concentration of oxygen bound by leghaemoglobin in the nodule is of the order of air-saturated water and that the concentration of free oxygen has been reduced to a very small fraction of it.

Proportion of leghaemoglobin oxygenated	Concentration of free oxygen (moles l^{-1})	Concentration of bound oxygen* (moles l^{-1})
0.95	7.1×10^{-7}	6.5×10^{-4}
0.50	3.7×10^{-8}	3.4×10^{-4}
0.20	9.4×10^{-9}	1.4×10^{-4}
0.05	2.0×10^{-9}	3.4×10^{-5}
Air-saturated water at 15°C	2.4×10^{-4}	

*Based on a total leghaemoglobin concentration of 6.8×10^{-4}M

concentration of leghaemoglobin in the nodule is about 7×10^{-4} M and, on average, 20% of it is bound with oxygen. The bound oxygen concentration will then be 1.5×10^{-4}M. This is very much higher than the concentration of free oxygen with which it is in equilibrium it (1×10^{-8} M). Thus, with the assumption that diffusion is to zero concentration, oxygen bound to leghaemoglobin will diffuse about 1000 times faster than free oxygen. These facts strongly suggest that one role of leghaemoglobin is that of facilitating the diffusion of oxygen.

If leghaemoglobin is 95% saturated with oxygen then the concentration of bound oxygen in the nodule will be 6.6×10^{-4} M and the concentration of free oxygen in equilibrium with this will be 7.7×10^{-7} M. Thus if the concentration of oxygen is increased from 20% saturation of leghaemoglobin to 95% saturation the total oxygen concentration, bound and unbound, will increase by 5.1×10^{-4}M and the free, unbound, oxygen concentration will increase by only 7×10^{-7}M, a difference of 700-fold. On the other hand if this increase were to be all free oxygen it would be in equilibrium with 7.6% oxygen in the gas phase, a concentration at which all the nitrogenase would be inactivated. It is not suggested that the average concentration of oxygen-saturated leghaemoglobin would be as high as 95% but it is clear that the nodule can hold a large amount of oxygen at a very low concentration of free oxygen. It would thus seem reasonable to conclude that, as well as serving to facilitate oxygen diffusion, leghaemoglobin can also serve as an oxygen buffer and while increasing its loading still maintain a low oxygen concentration at which the nitrogenase in the bacteroids can function.

The third function that was considered, that of conveying oxygen to specific sites on the bacteroid surface, is not likely as it has been shown that a number of other oxygen binding proteins such as myoglobin can substitute for leghaemoglobin with isolated bacteroids.

4.3 Hydrogen

In 1976 Schubert and Evans published work which showed that legume root nodules evolved a much greater amount of hydrogen than is consistent with an efficient nitrogenase system. They introduced the concept of Relative Efficiency, RE:

$$RE = 1 - \frac{\text{rate of } H_2 \text{ evolution in air}}{\text{rate of } H_2 \text{ evolution in 20\% } O_2, 80\% Ar}$$

(or rate of C_2H_2 reduction)

The rate of evolution of hydrogen under argon and oxygen, with no nitrogen present, gives the total activity of the enzyme. Thus the RE gives the proportion of the electrons which are donated to nitrogen. At full efficiency the RE should be 0.75 (see Chapter 3). Experimental values higher than this are the result of hydrogen recycling by uptake hydrogenase (see below). RE values as low as 0.4 were found, which indicated that 60% of the energy going to the nitrogenase system went to the production of hydrogen. This is very inefficient.

The question raised is of course why root nodules should be so inefficient. The answer has not yet been found to everyone's satisfaction. There is no reason to believe that the nitrogenase present in bacteroids is substantially different from the nitrogenase present in *Clostridium* and *Klebsiella*, and where they have been compared they have been found to be substantially the same. There is thus no reason to believe that the enzyme itself is less efficient. It must be some condition within the bacteroid which renders it so. In Chapter 3 some conditions in which excess hydrogen is evolved from nitrogenase were discussed. All but two of these conditions at the same time decreased the activity of the enzyme. It is clear that this is not the case in root nodules, as lower efficiency is more often correlated with higher nitrogenase activity.

The only two factors that accord with this are hydrogen inhibition and starvation of nitrogen as a substrate. Experiments in which higher pressures of nitrogen were used did not improve the efficiency of pea root nodules, and it has been calculated (Sinclair and Goudriaan, 1981) that nitrogen should diffuse into the nodule with sufficient rapidity to prevent this happening. We are left then, by a process of elimination, with hydrogen inhibition as being the most likely cause. Many people do not consider that this is possible because of the high diffusivity of hydrogen. We need to know the concentration of hydrogen within the nodule. Only when we know it will this cease to be an open question.

4.3.1 *Hydrogenase*

Many strains of *Rhizobium* produce the enzyme uptake hydrogenase when in bacteroid form. This enzyme is able to re-utilize the hydrogen evolved from nitrogenase. The roles of this enzyme were considered in Chapter 3. The hydrogenase is not able to reduce NAD^+ in experiments *in vitro*. It has been shown to reduce ubiquinone directly *in vivo*. If ubiquinone is the electron acceptor from uptake hydrogenase, then only some of the energy as ATP can be regained, as the ATP synthesis step between NADH and ubiquinone will be bypassed. If little energy is saved by the oxidation of hydrogen and little contribution can be made to respiratory protection, then the other function, that of decreasing hydrogen inhibition, would seem to be the only one left. This would be an important role if the theory that excess hydrogen evolution is due to hydrogen inhibition is correct.

Substantial improvements in yields of nitrogen fixed have been found by some workers when nodulated plants with and without hydrogenase have been compared. The strains of *Rhizobium* were different in these tests and the differences in yield could be due to the strain's other characteristics rather than to the presence of an uptake hydrogenase. This question, as to whether or not there are measurable gains to be made by using strains of *Rhizobium* that possess uptake hydrogenase is still not satisfactorily answered.

4.4 Carbon metabolism

Carbon is transported from the shoot to the root nodule in the form of sugars. In the nodule the sugars have to be transformed into the carbon skeletons of the nitrogen transport compounds (see Chapter 6), and also have to be respired to produce ATP and reducing power for nitrogen fixation. Although a lot of work has been done on the biochemistry of the root nodule, up until now most of this has been concentrated upon nitrogen metabolism rather than the carbon metabolism that supports it.

We want to know the identity of the compounds that reduce low-potential electron donors to nitrogenase and the metabolism that produces these compounds. We are, at the moment, in complete ignorance of this. We are, however, aware of some aspects of nodule carbon metabolism.

Do the bacteroids use sugars or are they first transformed to organic acids by the plant cytoplasm before being utilized by the bacteroids? Isolated bacteroids use dicarboxylic acids such as succinate and can fix nitrogen with them as substrates. Some genetic evidence supports the view that dicarboxylic acids are utilized by the bacteroids in the nodule. Mutants of *Rhizobium* that are unable to take up dicarboxylic acids are able to nodulate legumes quite efficiently but are unable to fix nitrogen. Similarly, mutants which lack succinic dehydrogenase are able to nodulate but not to fix nitrogen.

If bacteroids use dicarboxylic acids as their substrate to fix nitrogen then one of the roles of the plant cell will be to convert the sugars into these acids for use by the bacteroids. The normal metabolic route that comes to mind is the glycolytic pathway, followed by metabolism by the citric acid cycle to

BIOCHEMISTRY AND PHYSIOLOGY OF LEGUME ROOT NODULE 73

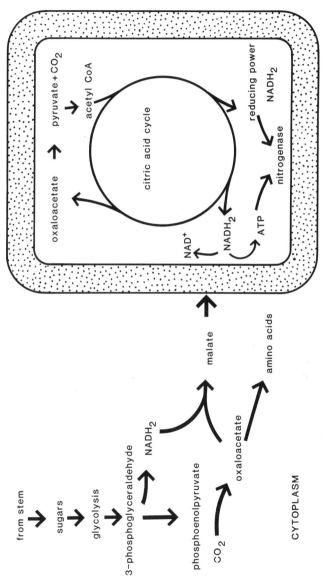

Figure 4.3 A scheme for the metabolism of carbon in pea root nodules.

produce succinate and malate. The citric acid cycle enzymes are present in the mitochondria. Removal of the intermediates in the cycle for use elsewhere would bring the cycle to a halt unless reactions occur which would replenish it (anaplerotic reactions).

There are, however, few mitochondria in the cell compared to the numbers of bacteroids, and there is thus some doubt whether they would be able to operate sufficiently rapidly to accomplish the transformation of sugars into organic acids at the necessary rates. However there is no need of the citric acid cycle to produce dicarboxylic acids. Legume root nodules possess the enzyme phosphoenolpyruvate carboxylase, PEP carboxylase:

$$\text{phosphoenolpyruvate} + HCO_3^- = \text{oxalacetate} + Pi$$

Thus sugars can be metabolized by the glycolytic pathway until phosphoenolpyruvate is produced which then can be carboxylated to oxalacetate. Some of this acid will be used to form aspartic acid and asparagine for nitrogen transport, but the rest will be reduced by malic dehydrogenase to malate. This can occur in the cytoplasm and does not need the intervention of the mitochondria (see Figure 4.3).

The bacteroids contain the citric acid cycle enzymes, so that metabolism of succinate will presumably go by that route. The pyruvate needed for the cycle to operate can come from the decarboxylation of oxalacetate. ATP is produced by oxidative phosphorylation. The identity of the reactions which produce the reductant for nitrogenase reduction are at present unknown.

The plasmalemma is not normally permeable to dicarboxylic acids, so that as the dicarboxylic acids have to pass through the peribacteroid membrane, which is derived from the plasmalemma, this membrane must be modified by incorporating a permease for these compounds.

Most of this chapter has been concerned with the diffusion of gases and the mechanisms that assist this. The root nodule has very high metabolic activity and the metabolic processes of nitrogen fixation and the respiration that supports it involve the uptake and evolution of gases. It is interesting to see how the nodule has evolved to deal with the problems that these diffusion processes pose. The problems of gaseous diffusion are not so acute in the actinorhizal root nodules, which are discussed in the next chapter.

CHAPTER FIVE

THE BIOCHEMISTRY AND PHYSIOLOGY OF ACTINORHIZAL NODULES

The development of techniques for isolation and growth of *Frankia* in culture has removed one of the major obstacles to the detailed study of nitrogen fixation and associated processes in actinorhizal nodules. Other problems which present difficulties in the study of the biochemistry of these nodules arise from their woody nature, high phenolic content and phenol oxidase activity, which make it difficult to prepare active enzyme extracts from nodule homogenates. These problems have been partly overcome by the extraction of nodules in strongly reducing media or under liquid nitrogen, and this has permitted the study of cell-free preparations of nitrogenase and hydrogenase and some of the host plant enzymes such as glutamine synthetase. The problem of pigmentation by phenolic oxidation products has, until recently, proved a major obstacle to the confirmation of the presence of haemoglobins in some actinorhizal nodules. As techniques for working with these nodules improve so we are becoming aware of the diversity of detail in the physiological and biochemical events which support nitrogen fixation in the wide range of plants that are nodulated by *Frankia*.

5.1 Gaseous diffusion

All aerobic nitrogen-fixing organisms possess some means of restricting oxygen diffusion as part of their defence mechanism against inactivation of nitrogenase. As discussed previously, in legumes the first barrier to the free diffusion of gases is the lack of intercellular spaces at the 'endodermal' tissue layer in the cortex. A similar barrier is also present in the nodule cortex of the *Rhizobium–Parasponia* association and restricts gaseous diffusion to the intercellular spaces in the bacterial zone which here is situated in the inner cortex, outside the centrally located stele. In actinorhizal nodules, although the central location of the stele in most species shows structural similarity to *Parasponia* nodules, the infected cells are grouped in longitudinal files in the cortex and there is no obvious barrier in this tissue to gaseous diffusion from the atmosphere. This is confirmed by observations of dyes infiltrated into the nodules. Diffusion occurs freely from the lenticels, through the large air spaces of the cortex to the air spaces which adjoin the infected cells.

Ethylene production from acetylene by *Alnus* shows diffusion-limited kinetics, so that barriers to the free diffusion of gases to nitrogenase do exist. There will be a restriction to diffusion by barriers such as the lipoidal fraction of the cell wall/plasmalemma of the infected cells or the wall of the endophyte vesicle. Suberization or lignification of the walls of infected cells has been reported in *Myrica*, *Alnus*, and *Casuarina* nodules, but whether this affects the diffusion of oxygen is not known. The appearance of nitrogenase activity in *Frankia* in nitrogen-free culture media is closely associated with the production of vesicles (Figure 5.1), and the occurrence of nitrogen fixation at atmospheric pO_2 (Figure 5.2) shows that some elements of vesicle structure must be of importance in protecting nitro-

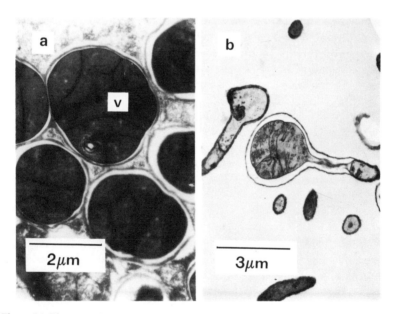

Figure 5.1 Electron micrographs of (*a*) vesicles, *v*, in the infected cells of a nodule of *Alnus glutinosa*, and (*b*) a vesicle from a nitrogen-fixing culture of *Frankia* 020607, isolated by Dr H.G. Diem from nodules of *Casuarina equisetifolia*. Septation of the vesicles can be seen in both cases.

genase from oxygen (Tjepkema *et al.*, 1981; Winship and Tjepkema, 1983).
A question, currently of topical concern, is the nature of the barrier(s) which regulate access. One obvious candidate is the multilaminate wall of these structures which bears many resemblances to the laminated glycolipid cell wall of the heterocysts in cyanobacteria. This, it has been suggested, may play a part in the control of oxygen penetration. While definite proof of such a role has not been obtained, similarities in the kinetics of oxygen and acetylene uptake by *Frankia* grown under nitrogen-fixing conditions and by heterocysts support the probability of a common mechanism for restriction of oxygen access. In both *Frankia* and heterocystous bacteria the K_m for acetylene is much higher for the intact organism than for nitrogenase *in vitro*, indicating that the intracellular levels of acetylene may be much lower than the external concentration, due to restriction of diffusion. In addition, vesicular (nitrogen-fixing) *Frankia* and heterocystous cyanobacteria both show diffusion-limited kinetics of respiration. The biphasic response of respiration to oxygen concentration in nitrogen-fixing *Frankia* is illustrated in the Lineweaver–Burke plots of Figure 5.3. Ammonium-grown cells show only one apparent K_m for O_2 of about 1 μM, which is similar to that of nitrogen-fixing cells at lower oxygen concentrations. The second kinetic component of oxygen uptake in nitrogen-fixing cells shows a linear increase in respiration with oxygen concentrations above 37μM and presumably derives from the special nature of the vesicles. Further study of the chemistry and physics of the vesicle cell wall structure are required to resolve precisely how they may function in regulating gaseous diffusion (Murray *et al.*, 1984).

Nitrogenase activity in vesicle clusters isolated from nodules has usually been detected only in preparations incubated anaerobically in the presence of ATP, with dithionite as the electron donor. Carbon substrates such as succinate have only occasionally been demonstrated to be taken up and respired aerobically at low rates by isolated vesicle clusters. It seems probable that during isolation, either mechanical or chemical damage (through the toxic effects of quinones formed from plant phenolics) to vesicle membranes is responsible for the oxygen sensitivity of such preparations. Vesicles isolated from *Frankia*, grown *in vitro*, will also only fix nitrogen under anaerobic conditions when supplied with reductant and an ATP-generating system. Mechanical damage again presumably permits access of oxygen to oxygen-sensitive sites which must be inaccessible in the intact organism (Akkermans *et al.*, 1983).

The relatively high solubility of carbon dioxide in both aqueous and organic solvents suggests that respired CO_2 should diffuse away relatively

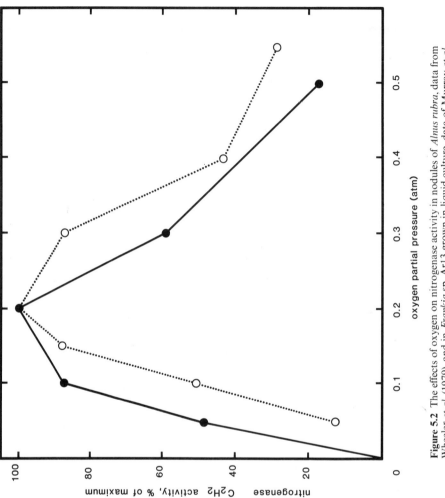

Figure 5.2 The effects of oxygen on nitrogenase activity in nodules of *Alnus rubra*, data from Wheeler *et al.* (1979), and in *Frankia* sp. Ar13 grown in liquid culture, data of Murray *et al.* (1984). ●, *Frankia* culture; ○, *Alnus* nodules.

BIOCHEMISTRY AND PHYSIOLOGY OF ACTINORHIZAL NODULES

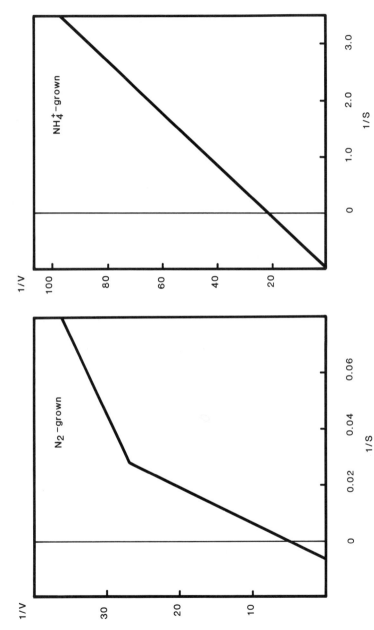

Figure 5.3 Reciprocal plots of oxygen uptake (μM) against oxygen concentration (n moles/mg protein/min) in the medium for ammonium-grown and nitrogen-fixing cultures of *Frankia* sp. Ar13: Adapted from Murray *et al.* (1984).

freely from the infected cells into the intercellular spaces of the nodules and thence to the atmosphere. It is not known whether actinorhizal nodules contain carbonic anhydrase which, it has been suggested, by catalysing the hydration of CO_2 to HCO_3^- may lower CO_2 tensions in the infected cells of legumes, as was discussed in the last chapter.

5.2 Hydrogen uptake and evolution

Unlike legumes, actinorhizal nodules have usually been found to evolve little hydrogen when active in nitrogen fixation, and will also consume added hydrogen. This process is oxygen-dependent, indicating the functioning of the oxyhydrogen reaction.

$$H_2 \rightarrow 2H^+ + 2e$$
$$2H^+ + 2e + 0.5\,O_2 \rightarrow H_2O$$

Studies with alder nodules have shown that uptake hydrogenase activity varies seasonally; although little hydrogen is evolved during the summer, in the autumn nodules which now fix nitrogen only at a slow rate show net hydrogen evolution and do not take up added hydrogen. The reasons for the autumn decline in uptake hydrogenase activity are not known. It is possible, however, that uptake hydrogenase may be of advantage to actinorhizal species in any soils where microbial activity liberates hydrogen, which might then be utilized via the oxyhydrogen action to provide additional ATP and reductant for the microsymbiont (Roelofsen and Akkermans, 1979).

Frankia hydrogenase has not yet been obtained in pure form. There are, however, some indications from the work of Benson, Arp and Burris (1980) with endophyte preparations that the hydrogenase systems may not be identical in all actinorhizal nodules. Vesicle cluster preparations from *Myrica pensylvanica* nodules showed hydrogen uptake and hydrogen evolving activity in the presence of oxidized and reduced viologen dyes respectively, whereas *Alnus rubra* endophyte showed only hydrogen uptake activity. There are various possible explanations for these differences: *Myrica* hydrogenase may be quite different from that of alder; the *Myrica* endophyte may contain two hydrogenases, a unidirectional uptake hydrogenase and a reversible type; or the differences could simply be due to greater susceptibility of the alder endophyte hydrogenase system to damage, so that the hydrogen-evolving activity was lost during extraction from the nodules. Further study of hydrogenase isolated from nitrogen-fixing cultures of *Frankia* strains from these two plant species may help to discriminate between these possibilities.

5.3 Haemoglobin

Spectroscopic evidence for the occurrence of haemoglobin in actinorhizal nodules was first provided by Davenport (1960), who observed absorption bands at 580 and 544 nm, resembling those of oxyhaemoglobin, in fresh preparations of *Casuarina* nodules. These bands were replaced by a single band at 562 nm, similar to that of deoxygenated haemoglobin, when these preparations were reduced with dithionite. The high phenol oxidase activity of actinorhizal nodule tissue soon discolours preparations, rendering spectroscopic examination rather difficult. It is only recently, therefore, that Tjepkema was able to confirm Davenport's results by spectroscopic observation of slices of root nodules of *Casuarina cunninghamiana* and *Myrica gale* (Figure 5.4). The concentration of haemoglobin in these two species was considered to be similar to that in *Pisum sativum* nodules, but lower concentrations were present in nodules of *Comptonia*, *Alnus* and *Elaeagnus*, while little or no haemoglobin was detected in *Ceanothus* and *Datisca*. There is currently no plausible explanation for these differences between species. Indeed, although the haemoglobin occurs in the nodules in a soluble form, certainly in *Casuarina* it is not clear whether it plays any role in the oxygen relations of actinorhizal nodules. While it is conceivable that, as in the nodules of legumes and of *Parasponia*, it may facilitate oxygen flux through the cytoplasm of the infected cell, equally it may be important in quite different ways, for example as a reserve of iron. Comparative study of the physiology of actinorhizal nodules with and without haemoglobin may help to shed further light on the functions of this pigment.

5.4 Carbon metabolism

Isolates of *Frankia* from a number of actinorhizal species have been divided into two broad groups which differ serologically in the sugar patterns revealed after whole cell hydrolysis and in their ability to utilize carbohydrates for growth (Lechevalier *et al.*, 1982). Group A strains comprise a heterogeneous group which are ineffective in the host plant. They have pigmented cells, diverse whole cell sugar patterns, based on a variety of sugars such as fucose and madurose, and can grow on media which contain carbohydrates such as glucose, maltose and xylose. Group B strains all have a whole cell sugar pattern which contains xylose. They are microae-

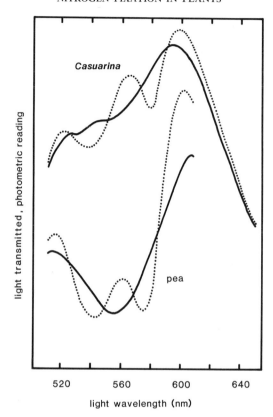

Figure 5.4 Absorption spectra of oxygenated and deoxygenated nodule sections of *Casuarina cunninghamiana* and *Pisum sativum* showing the presence of absorption bands attributable to haemoglobin. Measurements are of light transmitted, hence absorption peaks appear as minima in spectra obtained under N_2 (solid lines) or O_2 (dotted lines). Adapted from Tjepkema (1983).

rophilic and most do not utilize carbohydrates. Many of the isolates from *Alnus*, some *Myrica* isolates and isolates from *Comptonia peregrina* fit into this group. The more detailed studies of carbon metabolism are confined to group B strains at present. These strains generally grow well in liquid media which contain propionate as the carbon source. Activity of the enzyme propionyl CoA carboxylase has been detected in *Frankia*, the growth of which is stimulated by the carboxylase enzyme cofactor, biotin. The carboxylation of propionate leads to the formation of methyl malonyl CoA. Further metabolism of this probably involves isomerization to succinyl CoA in a reaction which is catalysed by the enzyme methyl malonyl CoA

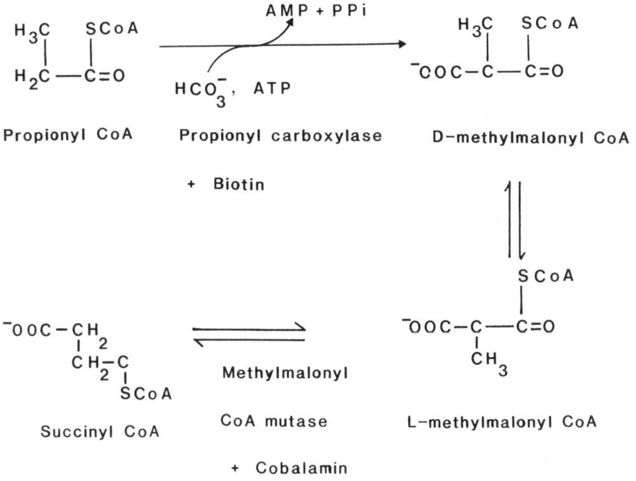

Figure 5.5 A possible route for the generation of dicarboxylic acid from propionate in *Frankia*.

mutase which requires Vitamin B 12 as a cofactor (Figure 5.5).

Differences in the ability of *Frankia* strains to utilize particular substrates are apparent from the differences in carbon substrates which will support the development of nitrogenase activity in some strains. When grown on nitrogen-free media, *Frankia* sp. AvcI 1 will utilize propionate, fumarate and malate as substrates for the development of nitrogenase, but other related organic acids are ineffective. *Frankia* sp. CpI 1 develops nitrogenase with succinate, malate or fumarate, but other carbon sources are ineffective. The full extent of the differences between strains and the precise metabolic routes for the respiration of the different substrates are currently unknown. *Frankia* sp. AvcI 1 shows activity of the glyoxylate enzymes, isocitrate lyase and malate synthase, when grown on acetate and fatty acids, so that these compounds could be converted into succinate for further metabolism. The inability of AvcI1 to utilize succinate for growth when supplied in the culture medium could be due to the loss or absence of an appropriate uptake system from this strain (Tjepkema *et al.*, 1981; Akkermans *et al.*, 1983). While the available experimental evidence points to the dicarboxylic acids as the substrates of respiration in cultured *Frankia*, unequivocal demonstration that these are the immediate carbon and energy source for *Frankia in vivo* has not yet been obtained. The close association of host plant mitochondria with *Frankia* in the infected cells creates considerable difficulty in proving that metabolic activity associated with isolated vesicle clusters is due to *Frankia* rather than to remnants of host cell cytoplasm. Thus Akkermans and co-workers observed partly disrupted mitochondria embedded in vesicle cluster preparations from *Datisca* nodules which might have been responsible for the relatively high rate of respiration of succinate. In preparations from *Alnus* and *Hippophae*, oxygen uptake by vesicle preparations was little stimulated by succinate. However, citric cycle enzymes were detected in these preparations, whereas enzymes of carbohydrate metabolism, such as hexokinase, pyruvate kinase and pyruvate dehydrogenase, and glyoxylate enzymes were not detected. It is possible, therefore that the transport system responsible for uptake of organic dicarboxylic acids by some *Frankia* strains *in vivo* may be disrupted during isolation.

As in legume nodules, additional input of organic carbon is provided during nitrogen fixation by carboxylation by the host plant cell of phosphoenol pyruvate to produce oxaloacetate. Study of the fate of the carbon from $^{14}CO_2$ fixation in alder nodules has shown that a large fraction accumulates as malate, formed by reduction of oxaloacetate, much of which is apparently not utilized in nodule respiration but may be exported from the nodule in the xylem as a counter-ion for cation transport. A smaller malate pool within the nodule is probably utilized in amino acid biosynthesis via the reactions of the citric acid cycle. In alders, the carboxylation reaction catalysed by carbamyl phosphate synthetase which produces carbamyl phosphate as a precursor of citrulline biosynthesis is another major route for dark CO_2 fixation (McClure *et al.*, 1983).

Different *Frankia* strains can vary considerably in their efficacy of nitrogen fixation in a particular plant species. Although the causes of such variation are obscure, variation in the effectiveness of symbiotic nitrogen fixation has been linked to the sporulating character *in vivo* of different *Frankia* strains (see Chapter 2). Norman and Lalonde (1982), showed for a range of isolates from *Alnus crispa* and *A. rugosa* that spore (−) isolates generally were more effective in symbiotic nitrogen fixation than spore (+) isolates. In a detailed study of spore (−) and spore (+) nodules of *Comptonia peregrina* and *Myrica gale*, VandenBosch and Torrey (1984) showed that both types of nodules evolved only small amounts of hydrogen and were thus equally efficient in recycling electrons lost as hydrogen from nitrogenase. However, the respiratory costs of nitrogen fixation were significantly higher in spore (+) than spore (−) nodules. The ratio of respiration (CO_2 evolved) to nitrogen fixed (C_2H_2 reduced) in spore (−) nodules was in the range 5.0 to 7.5, while for spore (+) nodules it was 10.0 to 11.7. The expenditure of photosynthate per unit of nitrogen fixed is thus

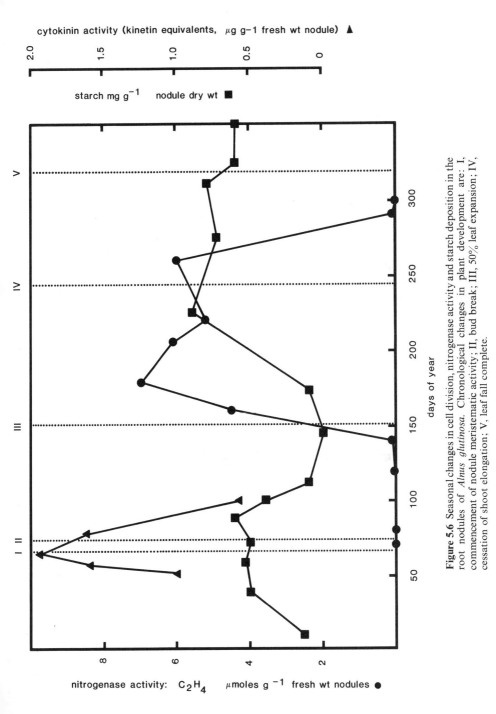

Figure 5.6 Seasonal changes in cell division, nitrogenase activity and starch deposition in the root nodules of *Alnus glutinosa*. Chronological changes in plant development are: I, commencement of nodule meristematic activity; II, bud break; III, 50% leaf expansion; IV, cessation of shoot elongation; V, leaf fall complete.

substantially higher in spore (+) than in spore (−) nodules. Sporangium differentiation within the nodule is therefore an expensive process in energetic terms. The only advantage possibly is to help to ensure the survival of the endophyte when it is liberated into the soil.

Little attention has been paid as yet to the formation of carbon reserve compounds in actinorhizal nodules, even though this clearly must be of considerable importance, in particular for overwintering of the perennial nodules of deciduous actinorhizal nodulated species. In alder, starch reserves are laid down in the nodule commencing in the summer, maximum accumulation coinciding with cessation of elongation growth of the shoot and a decline in nitrogenase activity of the nodules. These reserves are used to support nodule respiration during the winter, when nitrogenase activity normally cannot be detected. Interestingly, in the spring renewed meristematic activity and low levels of nitrogen fixation occur in the nodules about bud burst, before development of a new photosynthetic surface. Recommencement of cell division coincides with an increase in the levels of activity of cytokinins (plant cell division factors) in the nodules. Nitrogenase activity remains low, possibly suppressed by the accumulation of nitrogenous substance in the nodules, until leaves are about 50% expanded. Increased xylem transport, supported by leaf transpiration, then re-establishes the translocation system between nodule and shoot, and nitrogenase activity rapidly increases (Figure 5.6). Clearly, whatever photosynthates are available during early leaf expansion are utilized to support growth of the host plant rather than the energy-demanding nitrogen-fixing processes. As in all mutualistic symbiotic processes, the plant controls and limits to its own advantage the activities of the microsymbiont, which without controls would become parasitic on the host.

Finally, it should be stressed that our knowledge of the physiology and biochemistry of *Frankia* is based on the study of rather few strains. Recent studies of *Frankia* sp. ArI3 by Lopez and Torrey have shown that this strain has a fully operational glycolytic pathway, apparently unlike the Frankiae studied earlier by Akkermans and co-workers. Additionally, ArI3 has a pathway for gluconeogenesis and can synthesize large quantities of trehalose and glycogen when grown with propionate as carbon source. These observations demonstrate clearly the need for further experimentation with a range of strains to explore the biochemical properties of this new and exciting group of microorganisms.

CHAPTER SIX

NITROGEN ASSIMILATION

Ammonium, the first free product of nitrogenase action, is a toxic compound with a range of effects on metabolism, such as uncoupling oxidative phosphorylation or competing with the cationic cofactors of certain enzymes. The accumulation of ammonium in the microbial cell during nitrogen fixation must therefore be prevented. This is achieved in free-living nitrogen fixers by rapid utilization for growth and in symbiotic organisms by excretion into the host plant cell cytoplasm where it is assimilated into organic form. It is in the form of amino acids, amides or ureides that fixed nitrogen is made available to the host plant for its growth. The controls which are exerted on nitrogenase synthesis and activity by mineral nitrogen and the different pathways by which nitrogen is assimilated into organic combination are discussed in this chapter.

6.1 Regulation of nitrogenase by combined nitrogen

Under natural conditions, ammonium and nitrate are the forms of mineral nitrogen which are most likely to be available to support microbial and plant growth. With the notable exception of *Rhizobium*, which will be discussed later, uptake of either of these ions can repress both the synthesis and activity of nitrogenase. This is done by somewhat different routes depending on the ion which is accumulated. Studies with bacteria such as *Klebsiella* or *Azotobacter* have shown that the effect of ammonium on nitrogenase synthesis cannot be explained by a mechanism involving direct repression of the nitrogen-fixing (*nif*) genes. For ammonium to act as a repressor the cell must be capable of ammonium accumulation. If this is prevented—for example by treating cultures with methionine sulphoximine, a potent inhibitor of the primary ammonium-assimilating enzyme, glutamine synthetase—then nitrogenase synthesis is not repressed by ammonium. Nitrogenase synthesis in mutants which lack the ability to produce glutamine synthetase is also unaffected by ammonium. Although it was initially thought that glutamine synthetase itself might be directly implicated as a regulatory element for nitrogenase synthesis, further genetic studies now suggest that it is some other, as yet uncharacterized, products of the gene system coding for glutamine synthetase production which, by interacting with the regulatory genes in the *nif* gene cluster, switch *nif* gene expression on or off. In addition to this regulation of the transcription of the *nif* genes, ammonium has also been shown to destabilize *Klebsiella* mRNA, thus diminishing the coding capacity of the cell for nitrogenase (Eady *et al.*, 1981).

Nitrate also represses nitrogenase synthesis, not as might be expected by a direct effect on the *nif* genes or by its reduction to ammonium, but through its reduction to the intermediate product nitrite by the action of nitrate reductase. Evidence for this comes first from observations of the failure of nitrate to repress chlorate-resistant mutants of diazotrophs, which do not show nitrate reductase activity. Nitrogenase synthesis is, however, repressed by nitrite in these mutants. Secondly, Hom *et al.* (1980) monitored the rate of nitrogenase synthesis in *Klebsiella* by following the appearance of radioactivity in nitrogenase polypeptides after pulse feeding ^{14}C-labelled amino acids, and found that nitrogenase synthesis was not repressed by nitrate immediately, but that repression coincided with the appearance of nitrite in the growth medium. Repression by nitrate clearly does not involve the same *nif* regulator genes as ammonium repression, but seems to be linked to a redox change in the cell which results from nitrate reduction. The mechanism of nitrate repression of nitrogenase synthesis may be similar to the regulatory effects which oxygen has on the *nif* gene cluster, since nitrate can act as an alternative terminal electron acceptor to oxygen (Eady *et al.*, 1978). The regulation of nitrogenase synthesis by combined nitrogen at the genetic level is considered further in Chapter 7.

The activity of nitrogenase is not affected directly by nitrate and those effects which have been observed can be ascribed to damage of nitrogenase by nitrite. On the other hand, the addition of ammonium often inhibits the nitrogenase activity of many diazotrophs, the degree of inhibition being dependent upon growth conditions such as oxygen supply and pH. The enzyme is not subject to control by allosteric feedback inhibition since ammonium does not affect the activity of pure preparations of nitrogenase *in vitro*. Studies by Laane *et al.* (1980) showed that ammonium dissipates the electrical gradient, or membrane potential, across the cell membrane of free-living aerobic diazotrophs such as *Azotobacter*, and that it is this

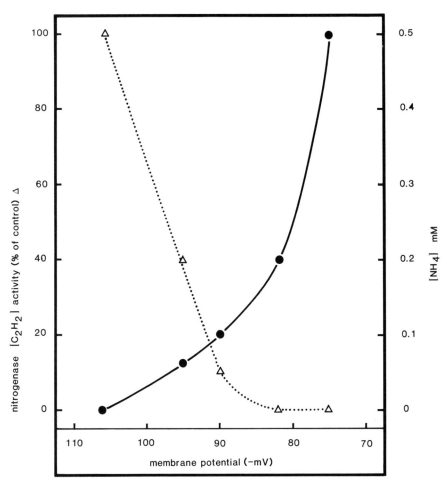

Figure 6.1 Dissipation of the electrochemical cytoplasmic membrane potential in *Azotobacter vinelandii* by ammonium and relationship to nitrogenase activity. Redrawn from Veeger *et al.* (1981).

which results in the switching off of nitrogenase activity (Figure 6.1). In *Azotobacter*, nitrogenase is inactive below a membrane potential of -80 mV. Reducing the membrane potential inhibits the generation of low-potential reducing equivalents and hence electron donation to nitrogenase. The ratio of ATP:ADP in ammonium-fed cells is largely unchanged from that in nitrogen-fixing cells so that ATP availability is not of immediate effect upon nitrogenase activity.

In the purple photosynthetic bacteria (Rhodospirillaceae) short-term inactivation of nitrogenase by ammonium is effected by an additional mechanism involving covalent modification of the Fe protein. Removal or addition of an adenine group inactivates or activates the Fe protein. The activating factor is a Mn^{2+}-dependent protein which is oxygen-sensitive (half-life of 2 to 3 minutes). The mechanism of this regulatory process is not yet fully understood, but it achieves an almost immediate switching on or off of nitrogenase in response to even small amounts of ammonium.

As has been indicated earlier, the effects of combined nitrogen on *Rhizobium* are rather different from those on free-living diazotrophs. Ammonium does not repress nitrogenase in strains that can fix nitrogen in culture, nor is there a direct effect on nitrogenase in bacteroid preparations. Inhibitory effects of ammonium arise indirectly and can usually be alleviated by manipulating the carbon or energy supplies of the system under investigation. The reason for this is that nitrogen-fixing *Rhizobium* cells do not accumulate ammonium. That which is formed by nitrogenase action is excreted, and ammonium supplied externally is not taken up by the cell. Under anaerobic conditions, Laane *et al.* (1980) found it possible to load *Rhizobium* bacteroids with ammonium chloride. This was because the pumping mechanism to exclude ammonium ions is dependent upon aerobic metabolism. Nitrogen fixation in these cells was inhibited by 80% over the controls when they were returned to aerobic conditions. This observation, together with the finding that treatment of bacteroids with the membrane probe tetraphenyl phosphonium bromide, which dissipates the membrane potential at the same time as inhibiting nitrogen fixation, suggests that the reducing equivalents required for nitrogen fixation are generated by the membrane potential as in some other diazotrophs. Nitrate also inhibits nitrogenase activity in free-living *Rhizobium*. The causative agent here is also nitrite produced by nitrate reductase activity, since nitrate does not affect nitrogenase activity in mutants which lack nitrate reductase, and nitrogenase is still synthesized in nitrate-treated cultures in which nitrogen fixation is suppressed by nitrate.

Nitrogen fixation in *Frankia* and in the heterocystous cyanobacteria requires the formation of specialized structures, vesicles and heterocysts (see Chapters 1 and 5), which provide a protected environment for nitrogenase action. As with other organisms, ammonium inhibits nitrogen fixation, and in cyanobacteria nitrate often has little effect. The formation of vesicles and of heterocysts is inhibited by the presence of ammonium, and in cyanobacteria such as *Anabaena cylindrica* the inhibitory effects of ammonium both on heterocyst formation and on nitrogen fixation can be prevented by the addition of methionine sulphoxime. It is probable that regulation of cyanobacterial *nif* genes involves a product of the glutamine synthetase gene system as in other aerobic diazotrophs.

The effects of combined nitrogen on the symbiotic association vary with both the host plant and the microsymbiont. Effects have been observed on virtually all stages of the root nodulation process from infection to the expression of nitrogenase activity and nodule senescence. Nitrate is generally viewed as being more inhibitory to nodulation than ammonium, with the production of nitrite being held responsible for a variety of effects on the nitrogen-fixing system, such as the inactivation of leghaemoglobin or oxidation of indole-3-acetic acid possibly involved in nodule development. Inhibitory effects are localized primarily at the region of ion uptake. This is seen most clearly in the tropical legume *Sesbania rostrata*, which bears nodules on both stem and roots. Feeding of ammonium nitrate via the roots supresses root nodulation, but stem nodulation and nitrogenase activity are not inhibited.

It is probable that the inhibitory effects of combined nitrogen stem from a whole complex of events. The cell membrane potentials of both host plant and endophyte must be affected and other effects such as diversion of photosynthate away from the nodules to support uptake and assimilation of combined nitrogen by the root system are to be expected. This last effect can at best only partially explain the inhibitory effects of combined nitrogen, since there are many instances where measures taken to enhance the availablity of photosynthates within the host plant fail to reverse inhibition of nitrogen fixation. For example, Huss-Danell *et al.* (1982) showed that when alders were allowed to photosynthesize in $^{14}CO_2$, the percentage distribution of whole plant ^{14}C to the nodule was little affected when nitrogen-fixing plants were fed ammonium. The vesicles of *Frankia* showed considerable damage within four days of feeding ammonium, and the recovery of nitrogenase activity following the removal of ammonium salts was largely dependent upon the formation of new vesicles (Figure 6.2).

Although combined nitrogen above certain concentrations depresses symbiotic nitrogen fixation, the growth of plants supplied with adequate combined nitrogen is normally increased compared with plants solely reliant upon nitrogen fixation. Possible reasons for this are the greater energy requirements for nodulation and nitrogen fixation compared with the requirements for assimilation of combined nitrogen such as ammonium. Additionally, it is likely that there are greater constraints on the supply of reduced nitrogen from nodules distributed in a relatively localized manner over the root system compared with the uptake of mineral nitrogen

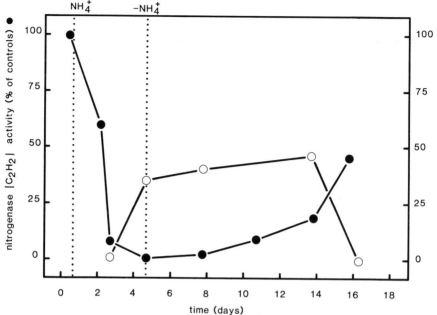

Figure 6.2 The effects of ammonium on nitrogenase activity and vesicle integrity in the nodules of *Alnus incana*. NH$_4$Cl (20 mM) was supplied to the plants in the nutrient solution and removed as indicated. Redrawn from Huss–Danell et al. (1982).

by the more widely distributed surface of the root hairs. It is of interest that work by Ingestad (1980) with *Alnus incana* seedlings showed that some of the inhibitory effects of mineral nitrogen on nitrogen fixation can be avoided if the supply is gradually increased to match consumption over a period of time rather than as a single, higher concentration maintained at constant levels over the experimental period.

6.2 The primary reactions of ammonium assimilation

As we have seen, it is important that the concentration of ammonia formed by the nitrogenase reaction is kept low to prevent inhibition of the nitrogenase system. Glutamine synthetase is the enzyme concerned with the initial incorporation of ammonium into organic combination:

$$\text{glutamic acid} + NH_3 + ATP \rightarrow \text{glutamine} + ADP + Pi$$

Because it is coupled to the hydrolysis of ATP, the equilibrium of this reaction is well over to the right, and this together with the low K_m for ammonia, 0.02 mM, ensures that ammonia is assimilated quickly and that the concentration is kept low.

The nitrogen now incorporated into the amide group of glutamine must be transferred to become the α-amino nitrogen of the amino acids. This is achieved by a further reaction catalysed by glutamate synthase, also referred to as glutamine oxoglutarate aminotransferase (GOGAT):

$$\text{2-oxoglutarate} + \text{glutamine} + \text{NADPH} + \text{H}^+ = 2 \text{ glutamate} + \text{NADP}^+$$

This equation is very similar to that for the reaction catalysed by glutamic dehydrogenase:

$$\text{NH}_3 + \text{2-oxoglutarate} + \text{NAD(P)H} + \text{H}^+ = \text{glutamic acid} + \text{NADP}^+ + \text{H}_2\text{O} + \text{H}^+$$

The only difference between the glutamine synthetase–GOGAT system and glutamic dehydrogenase is that ATP is hydrolysed in the former. This extra energy is expended in order to keep ammonia low; glutamic dehydrogenase has a small equilibrium constant and is thus readily reversible. It also has a much higher K_m for ammonia, 10 mM, than glutamine synthetase, which means that it cannot lower the ammonium concentration sufficiently to prevent inhibition of nitrogenase synthesis. The glutamine synthetase GOGAT system is summarized in Figure 6.3. From this figure and from the equations it can be seen that glutamic acid and glutamine merely serve as carriers and holding compounds for nitrogen before it is incorporated into the α-amino nitrogen of other amino acids or used for the synthesis of other nitrogen compounds for transport. The compounds utilized for transport of nitrogen differ both in the energy expended during their biosynthesis and in the C:N ratio within the molecule. This has consequences for the overall metabolic cost of the method of nitrogen utilization used. This is illustrated in the following discussion of the biosynthesis of the main groups of nitrogen compounds which are utilized for the transport of fixed nitrogen in root-nodulated plants.

6.2.1 *Amides*

Glutamine synthetase activity is negligible in *Rhizobium* bacteroids and in *Frankia* vesicles, and the ammonia excreted during nitrogen fixation is assimilated into organic form by host plant enzymes. Glutamine, the first product of ammonium assimilation, is usually present in xylem sap collected from amide transporting plants but is not the major nitrogenous constituent. This role is assumed by asparagine, which is more soluble and

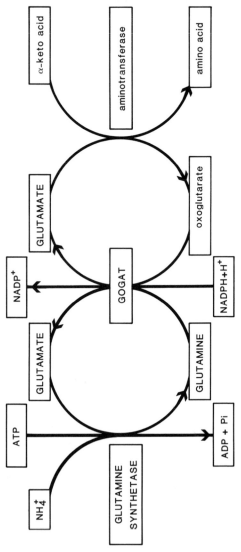

Figure 6.3 Reactions of the glutamine synthetase–GOGAT system for the assimilation of ammonium into amino acids. The overall equation for the reaction sequence is

$NH_4^+ + ATP + NADPH + H^+ + \alpha\text{-keto acid} = \text{amino acid} + NADP^+ + ADP + Pi + H^+$

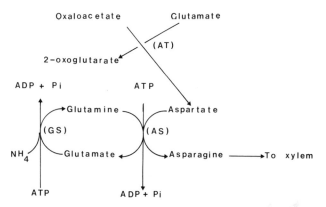

Figure 6.4 Metabolic flow chart for asparagine biosynthesis. Enzymes are indicated in parentheses: At, aminotransferase; GS, glutamine synthetase; AS, asparagine synthetase.

less metabolically active than glutamine. It has a more favourable C:N ratio of 4:2 (compared with 5:2 in glutamine) which effectively reduces the energy cost of nitrogen transport. Asparagine synthesis is catalysed by asparagine synthetase, an enzyme for which the K_m for glutamine is at least two orders lower than for ammonia. Consequently it utilizes the amide group of glutamine as the nitrogen donor in a reaction that requires ATP:

$$\text{glutamine} + \text{aspartate} + \text{ATP} \rightarrow \text{asparagine} + \text{glutamate} + \text{AMP} + \text{PPi}$$

Aspartate is generated in a transamination reaction between glutamate and oxaloacetate. It can be seen from the flow chart for asparagine biosynthesis (Figure 6.4) that the equivalent of seven ATPs are required for the synthesis of one asparagine molecule. An additional ATP is required to drive the anaplerotic reaction that incorporates CO_2 to replenish the citric acid cycle. Each carbon atom in asparagine represents a potential loss of six ATPs which could be obtained from its respiration by the citric acid cycle. The total cost of asparagine synthesis is therefore 32 ATP per asparagine molecule, or 16 ATP per ammonia assimilated. There is an additional, unquantified, expenditure of energy for the biosynthesis of the enzymes involved in amide synthesis and also in transport costs.

6.2.2 Ureides

The major derivatives of urea which have been found in nodulated plants are allantoin, allantoic acid and citrulline. The formulae of these com-

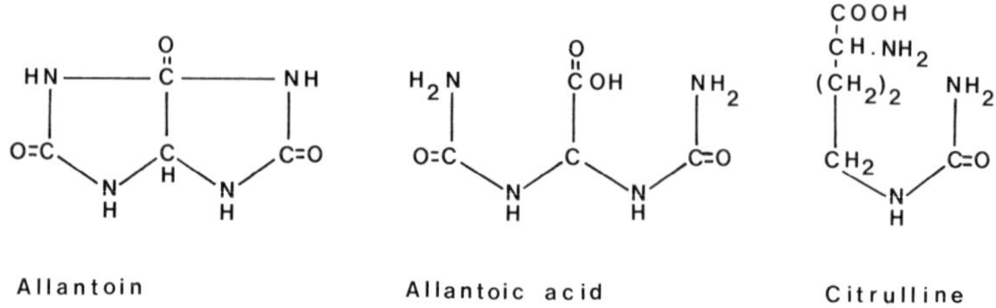

Figure 6.5 Structures of allantoin, allantoic acid and citrulline.

pounds are shown in Figure 6.5. The ratio of C:N in the first two compounds is 1.0 which provides a clear advantage over the amides in terms of carbon economy for nitrogen transport. The amides are the more common transport compounds in temperate legumes such as *Pisum*, *Vicia* or *Lupinus* and in most actinorhizal species. Allantoin and allantoic acid, the hydrolysis product of allantoin, are the favoured transport compounds of tropical and subtropical legumes such as *Glycine*, *Phaseolus* and *Vigna*. Citrulline is found as a major nitrogenous constituent of a smaller number of nodulated woody species. For example, it is present together with allantoin and allantoic acid in *Albizzia* xylem sap and it is a major amino acid in the nodules of *Alnus*.

Allantoin is the diureide of glyoxylic acid. However, comparison of the metabolism of ^{14}C-labelled glyoxylate and urea with that of glycine and the purine base, hypoxanthine, indicates that synthesis occurs not from urea and glyoxylate but via a pathway involving purine synthesis and degradation. The nucleotide first formed is inosine monophosphate (Figure 6.6). Studies by Atkins, Boland, Schubert, Hanks and others suggest that the synthesis of this compound occurs within, or is closely associated with, a nodule plastid fraction which can be isolated from nodule homogenates by centrifugation through sucrose density gradients. Some of the key enzyme activities of this pathway have been detected in this fraction, and radioactivity from ^{14}C-glycine is incorporated into purines when plastids are incubated with ribose-5-phosphate and other reactants involved in the synthesis, such as ATP, glutamine, aspartate, bicarbonate, methenyltetrahydrofolate, $MgCl_2$ and KCl. Further enzymic and radiotracer experiments have shown that glycine synthesis also occurs in the plastid fraction. The activity of phosphoribosyl pyrophosphate synthetase correlates with both nitrogenase activity and the rates of ureide export from soya bean nodules.

The major experiments which showed the involvement of purine degradative reactions in allantoin synthesis were carried out prior to the resolution of the steps leading to inosine monophosphate synthesis. In addition to the information gained from radiotracer experiments, it was found that feeding the xanthine dehydrogenase inhibitor, allopurinol, to nodulated soya bean or to cowpea resulted in the accumulation of the pathway intermediate, xanthine, in the nodules and a rapid decrease of the ureide end-products in nodules and stems, but not roots. Confirmation of the pathway has been obtained by a study of the enzymes involved which has demonstrated the occurrence of inosine monophosphate dehydrogenase, xanthine dehydrogenase, uricase and allantoinase in various

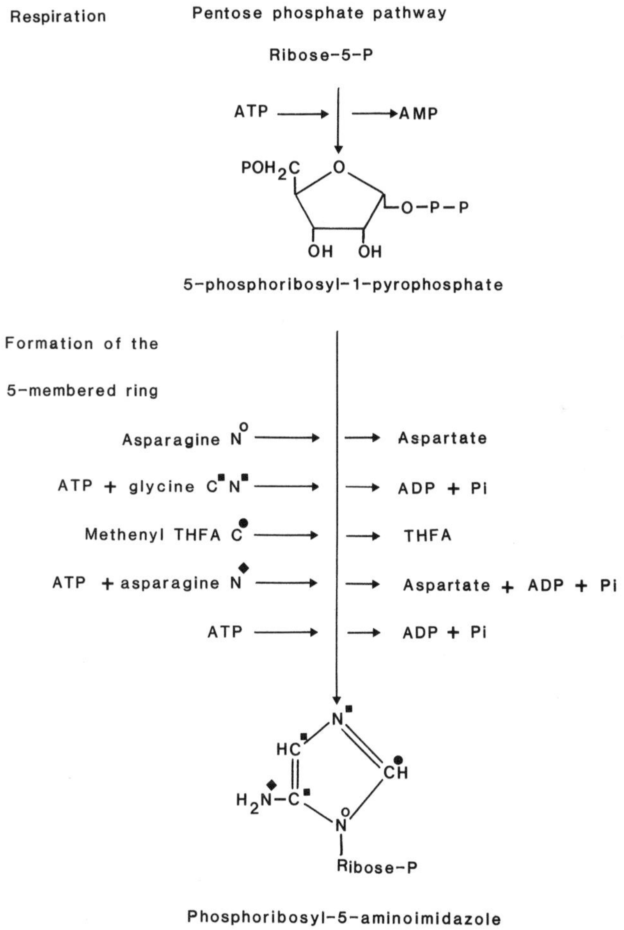

Respiration   Pentose phosphate pathway

Ribose-5-P

ATP ⟶ | ⟶ AMP

5-phosphoribosyl-1-pyrophosphate

Formation of the
5-membered ring

Asparagine N° ⟶ | ⟶ Aspartate

ATP + glycine C*N* ⟶ | ⟶ ADP + Pi

Methenyl THFA C• ⟶ | ⟶ THFA

ATP + asparagine N♦ ⟶ | ⟶ Aspartate + ADP + Pi

ATP ⟶ | ⟶ ADP + Pi

Phosphoribosyl-5-aminoimidazole   *Contd.*

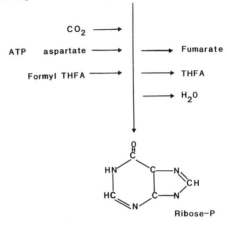

Figure 6.6 The biosynthesis of inosine monophosphate.

fractions of soya bean or cowpea nodules. The first enzyme is found in the proplastid fraction and xanthine dehydrogenase is located in the cytoplasm. The final oxidative reactions probably take place in the uninfected cells of the central nodule tissue in which higher activities of uricase are located in the peroxisomes and allantoinase is associated with the microsomal fraction. Uricase from cowpea and soya bean nodules has a high K_m for oxygen, so that the location of this enzyme in the uninfected cells may help to circumvent the oxygen limitation which could result from competition with bacteroid cytochromes for oxygen bound to leghaemoglobin in the infected cells. These events are summarized in Figure 6.7.

Calculation of the theoretical ATP cost for this pathway is more difficult than for the synthesis of amides. On the same basis as that employed for the calculation of the ATP cost for amide synthesis, Pate and Atkins (1983) estimate that 8.5 ATPs are required for each NH_3 assimilated. This is almost half the theoretical ATP requirement for the incorporation of NH_3 into asparagine. Ureide synthesis is favoured even more by the lower C:N ratio of allantoin and allantoic acid (1.0) compared to that of asparagine (2.0) with a concomitant reduction in carbon and nitrogen transport costs. However, the costs that are entailed in the synthesis of the

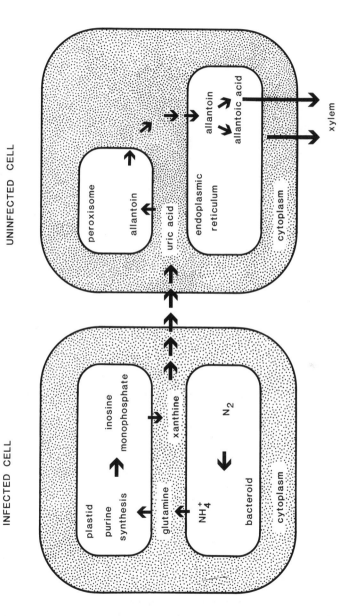

Figure 6.7 Location and organization of the pathway for ureide synthesis in legume root nodules. Adapted from Schubert and Boland (1984).

nitrogen compounds are not the only costs involved, and for the ureides there is unquantified energy expended in inter- and intracellular movement of pathway intermediates, as well as in the synthesis of the enzymes involved in the many steps of ureide biosynthesis and in the utilization of the nitrogen of these transport compounds when they are off-loaded from the xylem to support cellular biosynthesis.

Citrulline has been recognized as the major carrier of fixed nitrogen in alder for many years, but its synthesis has been less well studied. Cytoimmunochemical studies have shown that glutamine synthetase is located in the inner cortical cells of alder nodules but not in endophyte vesicles, and radiotracer studies with ^{13}N-labelled ammonium have shown that this enzyme is largely responsible for the primary assimilation of ammonium, although if ammonium is exogenously supplied, glutamate dehydrogenase plays a minor role in assimilation. These observations, together with the earlier finding of Leaf *et al.* (1958) that the carbamino nitrogen of citrulline was most heavily labelled with ^{15}N when nodules of *Alnus glutinosa* were exposed to ^{15}N-labelled N_2, suggest that the main path for assimilation of ammonia into citrulline is through the formation of carbamyl phosphate with glutamine as the nitrogen donor. Carbamyl phosphate synthetase is an unstable enzyme and has not been well characterized in plants; ornithine carbamyl transferase, which synthesizes citrulline from ornithine and carbamyl phosphate, occurs in two isozymic forms in the host plant cytoplasm. The metabolic pathway leading to citrulline biosynthesis is summarized in Figure 6.8.

Carbon dioxide fixation plays an important role in citrulline biosynthesis. Research by Schubert and co-workers has shown that approximately 40% of the $^{14}CO_2$ fixed by alder root nodules over a 20-minute period is incorporated into citrulline, with more than 80% of the ^{14}C located in the carbamyl group and thus assimilated during carbamyl phosphate synthesis. Phosphoenol pyruvate carboxylase activity is responsible for most of the remaining ^{14}C in citrulline. Most of this is located in the C-1 position and probably originated from glutamate. However, much of the CO_2 fixed accumulated in malate. Pulse labelling experiments suggested that there were at least two pools of malate in the nodules, one being rapidly converted into citrate which could be used to make glutamate while the second, larger pool, was metabolically inert. It was suggested that malate in this pool might be involved in pH control in the nodulated plant and transported from nodules as a counter-ion for cations.

Estimation of ATP consumption for citrulline synthesis suggests that an equivalent of 16 ATPs would be required for the synthesis of one molecule. An additional ATP would be required for phosphoenol pyruvate carbo-

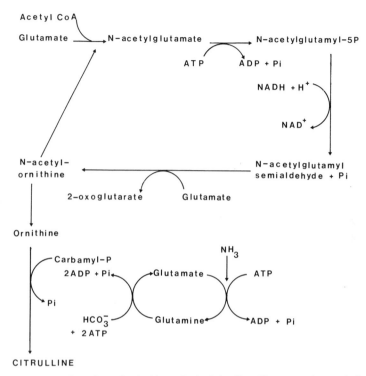

Figure 6.8 Metabolic pathway for the biosynthesis of citrulline. Glutamate nitrogen is derived from ammonia as shown in Figure 6.3. The acetyl group is first derived from acetyl CoA and can then be replaced by donation from N-acetyl ornithine in a cyclic reaction.

xylase activity. The five carbons of citrulline, other than that derived from carbamyl phosphate, represent a further potential loss of 30 ATPs, so that the total requirement in ATP equivalents for citrulline synthesis is 47 ATPs or 15.7 ATPs for each NH_3 assimilated. These costs are very similar to those for asparagine synthesis. Although the synthesis pathway to citrulline is more complex than that of asparagine, additional expenditure on the synthesis of appropriate enzymes will be offset by reduced nitrogen transport costs of the 3N citrulline molecule.

6.3 Transfer of fixed nitrogen in symbiotic systems

Symbiotic nitrogen fixation, whether by *Rhizobium*, *Frankia* or by cyanobacteria, occurs at localized sites within the host organism. Effective connections between the site of fixation and the body of the host are

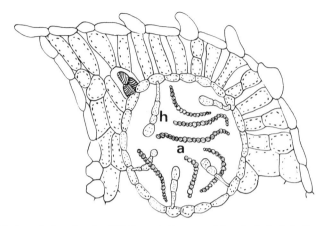

Figure 6.9 A vertical section (× 125) through the upper lobe of a leaf of *Azolla caroliniana* showing the mucilage-filled cavity lined with multicellular hairs, *h*, and containing filaments of *Anabaena, a*. Drawing kindly supplied by Professor G. Bond.

necessary to remove the products of nitrogen fixation rapidly. This is achieved in various ways in different symbiotic systems. In *Azolla* fronds the cyanobacterium *Anabaena* occupies a mucilage-filled cavity lined with multicellular hairs (Figure 6.9). These have wall infoldings typical of transfer cells and thus appear to be special adaptations for the transfer of fixed nitrogen to the host plant, although similar structures are also present in cavities of uninfected fronds when *Azolla* is grown on combined nitrogen (see also Chapter 2). The dense cytoplasm of the projecting cells is suggestive of intense metabolic activity. Presumably they are involved in the assimilation of ammonium excreted by *Anabaena* into the cavity. A rather similar arrangement is found in liverworts such as *Blasia* or *Anthoceros*, where the bryophyte produces branched filaments which project into the mucilaginous cavity in the thallus which contains *Nostoc*. These filaments also have wall infoldings typical of transfer cells and the structure again presumably functions to permit rapid movement of solutes between the cavity and the liverwort thallus. In *Azolla* assimilation of ammonium by the fern occurs via the glutamine synthetase–GOGAT pathway. However, in lichens such as *Peltigera aphthosa*, assimilation of ammonium excreted by the cyanobacterium, *Nostoc*, is by fungal glutamate dehydrogenase. Alanine is synthesized subsequently from glutamate by the aminotransferase reaction and it is in this form that the fixed nitrogen is transferred to the rest of the thallus.

Cyanobacterial root nodules are found among the cycads and also form

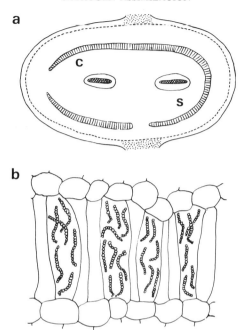

Figure 6.10(*a*) Drawing of a transverse section (× 12.5) of a branched nodule lobe from *Encephalartos villosus* root nodules. Two steles, *s*, can be seen in the branched structure. The cyanobacterial zone, *c*, is located in the mid cortex. (*b*). Detailed drawing (× 100) of part of the cyanobacterial zone of an *Encephalartos* nodule, showing filaments of the cyanobacterium occupying the mucilage-filled intercellular spaces. Drawing kindly supplied by Professor G. Bond.

an association with the angiosperm *Gunnera* (Chapter 2). In nodules such as those of *Encephalartos* or *Macrozamia* the cyanobacteria are located in mucilaginous spaces between the cells in the mid-cortex (Figure 6.10*a*, *b*). The coralloid nodules are of indeterminate growth and are well supplied with vascular connections back to the transport system of the host. In the *Macrozamia–Anabaena* symbiosis, citrulline and glutamine are the main nitrogenous compounds translocated (Halliday and Pate, 1976).

The process of nitrogenous solute transport has been examined most thoroughly in legumes largely as the result of the research of Pate and his co-workers. Species with nodules of determinate growth, belonging mainly to the Phaseoleae, have a closed vascular system, do not develop transfer cells in the pericycle, and export ureides. Nodules of other species that have indeterminate growth often develop pericycle transfer cells and export

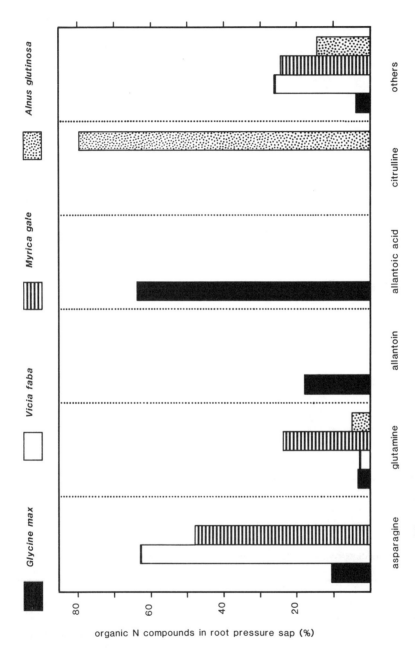

Figure 6.11 The composition of the bleeding sap of legume and actinorhizal root nodules.

amides. Transport of nitrogenous compounds probably follows a symplastic route via plasmodesmatal connections from the infected cells through the nodule cortex and endodermis to the pericycle which is thought to actively secrete compounds to the apoplast of the vascular tissue. In the 'open' vascular system of species such as *Pisum*, Minchin and Pate (1973) have calculated that half of the requirement for water for export of nodule nitrogenous solutes through the xylem may be derived from the input of phloem translocate into the nodule. Other contributions of water could come from osmotic movement from the parent root or, under conditions of ample water supply, from the rooting medium by absorption through the nodule epidermis. The transpiration stream is of major importance in the removal of products of nitrogen fixation from the nodules. In peas nitrogenous solutes have been observed to accumulate in the nodules at night or under conditions of high humidity during the day when shoot transpiration is low. These levels decrease rapidly as nitrogenous compounds are swept out of the nodules on resumption of transpiration. Transpirational flow may be of even more importance in the export of solutes from determinate nodules where resistance to water circulation should be reduced by the closed vascular system. Because ureides are of lower solubility than amides, a higher rate of water flow through determinate nodules could be of importance in keeping ureide concentrations in xylem sap below crystallization point (Pate, 1980).

Actinorhizal nodules are of indeterminate growth and differ from legume nodules in having a centrally situated stele, which is thus located inside rather than outside the infected tissues. Within the nodule the infected cells are arranged in longitudinal files radiating away from the stele. Groups of cells are separated from each other by files of uninfected cells. Microautoradiographs of actinorhizal nodules taken after feeding shoots $^{14}CO_2$ show much higher radioactivity in the infected rather than the uninfected cells, and it is tempting to suggest that these cells form the path for translocation of assimilates into and out of the nodules (Wheeler and Lawrie, 1976). Transfer cells have not been observed in actinorhizal nodules.

Much of the information concerning the nature of the nitrogenous substances exported from root nodules to the host plant has been obtained from the analysis of bleeding xylem sap collected from detached nodules or from the stumps of decapitated plants. As already indicated, such sap is enriched with particular components of the soluble nitrogen compounds in the nodules: the composition of bleeding sap from amide and ureide exporting legumes and actinorhizal nodulated species is compared in

Figure 6.11. Organic acids such as malate also figure prominently in the bleeding sap of such plants. These anions help to achieve ionic balance in xylem sap since plants fixing nitrogen tend to take up an excess of cations over anions from the soil solution.

Assimilation of inorganic nitrogen has different effects on the nature of the nitrogenous compounds transported in different groups of nodulated plants. Most relevant data comes from studies with legumes where the nitrogen source affects amide- and ureide-exporting species differently. In amide exporters such as pea or lupin, the distribution of nitrogen between the nitrogenous constituents of xylem sap is similar in both nitrate-fed and nitrogen-fixing plants. In ureide-exporting species such as soya bean or cowpea, however, the balance of nitrogenous constituents is changed so that nitrate-fed plants export more asparagine than the predominantly ureide-exporting nitrogen-fixing plants. The reduction of nitrate in nitrate-fed plants must occur mainly in uninfected regions of the root system, the cells of which presumably lack the regulatory mechanisms which induce increased activity of key enzymes of ureide synthesis during nodule development and nitrogen fixation. A lower level of synthetic activity allows the synthesis of a small amount of some ureides in root tissue however (Pate *et al.*, 1980).

The nitrogen fixed in root nodules is not of course all utilized for synthetic purposes elsewhere in the plant. Modelling techniques developed by Pate and co-workers suggest that up to half of the nitrogen required for nodule growth in nitrogen-fixing legumes may be obtained following xylem-to-phloem transfer of fixed nitrogen cycled through the shoots of the nodulated plant, while roots receive up to 90% of their nitrogen by this route. Mathematical modelling appears to offer an accurate method, based on available data, by which symbiotic nitrogen fixation can be integrated into and described in terms of whole plant economy. This is an approach which could be applied fruitfully to the symbiotic systems other than legumes as more basic experimental data become available.

CHAPTER SEVEN

THE GENETICS OF NITROGEN FIXATION

The control of metabolic processes is achieved in a variety of ways. Feedback inhibition and allosteric regulation of enzymes accomplish fine control of the actions of enzymes already synthesized. The processes leading up to enzyme synthesis must also be controlled, as the synthesis of proteins is expensive in terms of the energy and intermediates required. This is achieved by the regulation of transcription of DNA to form messenger RNA (mRNA), and at the level of the translation of mRNAs into proteins. The nature of these controls is investigated by genetic analysis. In order to understand how nitrogen fixation is controlled we need to know about the genes concerned with the synthesis of proteins that are required for the fixation of nitrogen and also about genes which control the expression of these genes.

The genes concerned with the nitrogen-fixing system, the *nif* genes, have been investigated most thoroughly in *Klebsiella*. This is because *Klebsiella* is closely related to *Escherichia coli*, the bacterium in which most of the techniques of gene and DNA manipulation have been worked out. In this chapter the genetics of the *Klebsiella* nitrogen-fixing system will be described and then some other nitrogen-fixing systems will also be discussed. Finally, some genetic aspects of the legume–*Rhizobium* symbiosis will be considered.

7.1 The genetics of the *Klebsiella* nitrogen-fixing system

7.1.1 *The* nif *genes*

If one gene codes for the synthesis of one polypeptide, a consideration of the biochemistry of nitrogen fixation immediately reveals that several genes will be necessary to code for the nitrogen-fixing system. The Fe protein is composed of two sub-units but, as each of these is the same, one gene will code for this protein. The MoFe protein has two different sub-units, each of which will require one gene. The molybdenum cofactor will require a gene, and further genes will be necessary to code for any special electron donors in the system. The genes that code for these proteins are all adjacent to one another on the *Klebsiella* chromosome. They lie between the *his* gene, which codes for histidine synthesis, and the *shi* A gene, which codes for the synthesis of shikimic acid, a precursor on the pathway of aromatic ring biosynthesis.

Research over a period has gradually increased the number of *nif* genes known to lie in this region, until now 17 genes have been recognized, accounting for all the DNA on this stretch of the *Klebsiella* chromosome. A list of the *nif* genes together with their functions is given in Table 7.1. It will be seen that the role of a number of *nif* genes is still unknown, so that yet

Table 7.1 Functions of the *nif* genes of *Klebsiella pneumoniae*

Gene	Function of the gene or gene product
H	Codes for the sub-unit of the Fe protein
D	Codes for the α-sub-unit of the FeMo protein
K	Codes for the β-sub-unit of the FeMo protein
M	Activation of the Fe protein
B	Involved in the synthesis and insertion of the iron molybdenum cofactor, FeMoco
N	As for B
E	As for B
V	As for B
F	Codes for a flavodoxin
J	Codes for a pyruvate : flavodoxin oxidoreductase
A	Codes for an activator molecule for the other *nif* genes
L	Codes for a repressor molecule for the other *nif* genes
Q	Possibly concerned with molybdenum uptake
S	Possibly concerned with processing the FeMo protein
U	As S
X	Unknown
Y	Unknown

Figure 7.1 Map of the *nif* region of *Klebsiella pneumoniae*. The arrows indicate the direction of transcription of the operons. Hatched, genes for FeMoco; stippled, genes for nitrogenase proteins.

more work is required before we fully understand this *nif* region. A map of the *nif* region is given in Figure 7.1. The region is composed of seven operons (see Figure 7.3).

The *nif* H, D, and K genes have been shown to code for the polypeptides of the Fe and MoFe proteins. These genes are readily identified by the lack of, or alteration of, the particular proteins in mutants. Mutations in the *nif* M gene result in an inactive Fe protein, so that the product of the *nif* M gene must be involved in modifying the protein in some way, perhaps incorporating the Fe–S cluster. Similarly, mutations in several genes affect the activity of the MoFe protein. Mutations of *nif* V give an altered substrate specificity. These mutants are unable to reduce N_2 but can reduce acetylene (Mclean and Dixon, 1981). Carbon monoxide, which does not inhibit hydrogen evolution from normal nitrogenase, inhibits hydrogen evolution from *nif* V mutants. When FeMoco was obtained from the *nif* V mutant protein and was combined with protein of a *nif* B mutant, a protein from which FeMoco is absent, the *nif* V$^-$ phenotype was obtained. However when FeMoco from a normal protein was added to the *nif* B mutant protein a normal protein resulted. Thus it was concluded that the *nif* V product modifies FeMoco in order to produce effective nitrogenase. From studies of *nif* V$^-$ mutants it has been concluded that FeMoco contains the binding site for N_2 and CO (Smith *et al.*, 1984).

Three other genes, *nif* B, N and E, have been identified with the synthesis of FeMoco, and the *nif* Q product's action is thought to be the acquisition of molybdenum. Thus five of the genes, *nif* Q, B, N, E and V, are connected with the synthesis of the molybdenum cofactor. The products of the genes *nif* S and U are thought to modify the MoFe protein, although there is no hard evidence for this as yet. If this is true, then nine genes are needed to produce the complete active MoFe protein.

Two genes are concerned with electron transport to nitrogenase: *nif* F and J. Extracts of mutants of both of these genes can fix nitrogen if they are provided with the artificial electron donor, sodium dithionite. Extracts of *nif* F$^-$ mutants can be rendered active by providing *Azotobacter* flavodoxin. It is thus assumed that the product of *nif* F is a flavodoxin. The *nif* J product has been shown to be the enzyme pyruvate:flavodoxin oxidoreductase, which catalyses the oxidation of pyruvate to produce reduced flavodoxin:

$$\text{pyruvate} + \text{CoA} + \text{flavodoxin}_{ox} \rightarrow \text{acetyl CoA} + \text{flavodoxin}_{red} + CO_2$$

This enzyme has been purified and it has been shown that, with flavodoxin and the pure enzyme in the reaction mixture, the reduction of nitrogenase and then acetylene can be achieved. When the flavodoxin from *Azotobacter* is used, the activity is one-third of that with the flavodoxin from *Klebsiella*, which demonstrates that there is some specificity for the reductant and that flavodoxins from different species may differ (Shah *et al.*, 1983).

The genes which control the synthesis of the nitrogenase proteins will be present in all the species that fix nitrogen. However, the genes concerned with electron transport will differ, as the provision of electrons depends upon the metabolism of the organism. *Clostridium*, for instance, uses ferredoxin as the donor of electrons derived from the metabolism of pyruvate. It is interesting to note that the genes *nif* F and J for electron transport to nitrogenase are transcribed in the opposite direction to the other *nif* genes (Figure 7.1).

The genes *nif* X and *nif* Y have been identified by means of their polypeptide products from cloned fragments of the *nif* region. The function of these genes has however not been established. The remaining two genes are *nif* A and *nif* L. The purpose of these genes is to control the expression of the other genes in the *nif* region.

7.1.2 *Regulation of the* nif *genes*

As nitrogen fixation is such an energy-demanding process it would not be advantageous for the organism to produce nitrogenase in an environment in which there is abundant fixed nitrogen. It would also be a disadvantage to synthesize nitrogenase proteins when oxygen is present, as this would inactivate the proteins as soon as they were synthesized. It is to be expected therefore that there will be controls which sense both nitrogen and oxygen levels and so prevent the synthesis of these proteins. Control at the transcriptional level will also save the synthesis of the messenger RNA and it is at this level that most of the control is achieved.

Although we now know a lot about the control of nitrogen fixation and nitrogen metabolism in *Klebsiella*, we do not yet know enough to be able to understand fully the operation of this complex process. It is thus only possible to draw outlines here and the reader is referred further to R.A. Dixon (1984).

Nitrogen fixation is one facet of nitrogen metabolism in these organisms and thus comes under the control of genes that regulate nitrogen metabolism as a whole. For instance under conditions of high concentrations of ammonia, *Klebsiella*, like a large number of genera of bacteria, switches from glutamine synthetase to glutamic dehydrogenase to assimilate ammonia. At the same time it will need to switch off the *nif* genes. It is thus economical to use the same control system for genes controlling both nitrogen assimilation and nitrogen fixation. The genes concerned with overall control are separate from the *nif* genes on the chromosome and are called *ntr* genes.

There are three of these *ntr* genes; *ntr* A, B and C. Of these, *ntr* A is located at a site distant from the other control genes, and *ntr* B and C are on the same operon as the gene which codes for the synthesis of glutamine synthetase, *gln* A (Figure 7.3). This fact led to some misinterpretation when studies in gene control were first initiated, as polar mutations in *gln* A did not synthesize either glutamine synthetase or the *nif* gene products. A polar mutation affects not only the gene in which it occurs but also the other genes in the direction of the transcription of the operon. Similarly, mutations in which the promoter of *gln* A was constitutive, so that glutamine synthetase was always present, made the *nif* genes insensitive to ammonia concentration. It was thus thought that glutamine synthetase had a control function. When it was realised that *ntr* B and C were control genes on the same operon as *gln* A, this misconception was removed.

The genes that are controlled by the products of the *ntr* genes are *nif* genes, *gln* A, *aut*, *hut*, and *put*. The last three genes code for the breakdown of the amino acids arginine, histidine and proline respectively. The product of the *ntr* A gene is necessary for the transcription of these genes as when *ntr* A deletions or mutations occur these genes are not transcribed. However, although the *ntr* A gene product is necessary for transcription, it is not controlled by nitrogen and its control function is not known. The *ntr* C product also regulates the transcription of genes within this system. Thus activation of the *nif* genes and the nitrogen assimilation genes requires both the *ntr* A and *ntr* C products.

Within the *nif* region *ntr* C controls only the transcription of the *nif* AL operon. If *nif* A expression is rendered constitutive then there is no requirement for the *ntr* C gene product for the activation of the *nif* genes. This is evidence for a cascade system whereby the *ntr* C and the *ntr* A products activate the transcription of *nif* A, which then produces a product

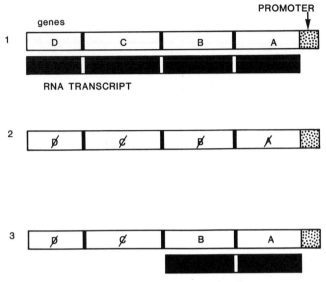

Figure 7.2 The function of operons. An operon is a transcription unit. All the genes on it are governed by the same promoter (see below). Mutations can have a polar effect. If a gene becomes inoperative such that the reading of the bases will be altered, all genes further along the operon are affected. Genes that are nearer to the promoter than the mutated gene are not affected.

1. If activated all the genes are transcribed.
2. If a mutation is in gene A or the promoter region no genes are transcribed.
3. If a mutation is in gene C then genes A and B are transcribed but not genes C and D.

A promoter is a region at the start of an operon which contains the transcription start site to which the RNA polymerase becomes attached and transcription can begin. It also contains sequences of bases to which activators or repressors can become attached, to either promote or hinder transcription. Control of genes on the operon is thus exerted here.

that will activate the other *nif* genes. The genes which require the *ntr* C product also require the *ntr* A product. Where the *ntr* C product is replaced by the *nif* A product in the *nif* genes, the *ntr* A product is still required. The control of the genes associated with nitrogen metabolism in *Klebsiella* is summarized in Table 7.2.

The positive control (activation) provided by the gene products of *ntr* A and C and *nif* A is not related to the amount of nitrogen or oxygen in the organism, so that other gene products need to be produced that can inhibit the transcription of the genes under conditions of high oxygen and combined nitrogen. The presence of oxygen or excess combined nitrogen

Table 7.2 The control of nitrogen assimilation in *Klebsiella pneumoniae*

Genes affected	High ammonia	Low ammonia	Very low ammonia
gln A, aut hut, put	− (ntr B + C)	+ (ntr A + C)	+ (ntr A + C)
nif AL	− (ntr B + C)	+ (ntr A + C)	+ (ntr A + C)
Other nif genes	− (ntr B + C)	− (nif A + L)	+ (ntr A + nif A)
Type of assimilation	Ammonia uptake GDH	Ammonia and amino acid uptake, GS + GOGAT	Nitrogen fixation GS + GOGAT

The promoters which allow the transcription of the genes are activated (+) or activation is inhibited (−) by the products of the genes that are indicated in brackets.

A combination of *ntr* A and C products activates, but when the product of *ntr* A combines with that of *ntr* B activation is inhibited. The products of *ntr* A and *nif* A activate the *nif* genes, but when the product of *nif* A combines with that of *nif* L, activation is inhibited.

GDH, glutamic dehydrogenase; GS, glutamine synthetase; GOGAT, glutamate synthase. For the genes see text.

---

must produce a signal that can be recognized so that control genes are able to operate. At the moment we do not know what these signals are. We can come a little closer in the case of nitrogen than with oxygen because there is some evidence that glutamine is involved. Some mutants have been found which produce defective glutamine synthetase constitutively and are thus unable to synthesize glutamine. These mutants also expressed the *nif* genes constitutively and the *nif* genes were not subject to ammonia control. This points to glutamine being concerned with the message that ammonia is high (Tubb, 1974).

Another class of mutants, defective in the *ntr* B gene, have been found in which the *nif* genes are not repressed by high levels of ammonia and which produce glutamine synthetase and glutamine. It is the product of the *ntr* B gene then that responds to the signals of high ammonia concentration, but the product of the *ntr* C gene is also required for it to act as a repressor. It is thought that the *ntr* B gene product interacts with the *ntr* C product so that when it binds to the promoter, the complex inhibits transcription of the operon to which it binds (Alvarez–Morales *et al.*, 1984).

The *ntr* B gene is not concerned with oxygen levels, as it controls genes that must operate to allow the assimilation of combined nitrogen when *Klebsiella* is living aerobically and when nitrogen fixation is not taking place. It is known that oxygen regulation is mediated by one of the *nif* gene products, that of *nif* L, since cultures which expressed the *nif* genes in the presence of oxygen were found to have a mutation in *nif* L.

The *nif* genes are more sensitive to the concentration of combined nitrogen than the other genes controlled by the *ntr* genes and are repressed by lower concentrations of nitrogen. Mutations in the *nif* L gene allowed expression of the *nif* genes at these lower concentrations of nitrogen. There is evidence that the *nif* L product acts by inactivating the product of *nif* A which is required for the transcription of the *nif* genes rather than by direct action on the individual operon promoters (Dixon *et al.*, 1983; Collins and Brill, 1985).

The consideration of the repression of the nitrogen-fixing genes makes the desirability of having a cascade system of control more apparent. At high combined nitrogen concentrations the organism does not need the more energy-demanding method of assimilating ammonia by glutamine synthetase or need to catabolize the amino acids, histidine, arginine and proline, and can (by repressing the necessary genes) save the cost of making the enzymes. Similarly, when ammonia is present at lower concentrations it can be assimilated using glutamine synthetase, with its lower K_m and also amino acids can be catabolized to obtain ammonia to synthesize such different amino acids as are required. Then only at very low ammonia concentrations will an organism switch on the nitrogen-fixing system which is so expensive in energy and protein. Up to 10% of the protein of the organism can be taken up by the nitrogen-fixing protein when the synthesis of these proteins is derepressed. The large amounts are required as the Fe and MoFe proteins have a low turnover number (a relatively low rate of catalysis).

7.2 The genetics of other nitrogen-fixing systems

The Fe and MoFe nitrogenase proteins are very similar in a number of organisms, so that the Fe protein from one species may be active in combination with the MoFe protein from another species, although this is not invariably so. One would expect therefore that the amino acid sequences would be closely similar and hence that the DNA coding for these proteins would have a high degree of homology. This has been found to be so, particularly for the *nif* H and D genes, and it has been possible to detect and map the corresponding genes in a number of organisms by finding regions of DNA homologous with the *Klebsiella* genes.

The metabolic environment of the nitrogen-fixing system is not the same in different types of organism, so that it would be expected that control of *nif* genes might differ. For instance, fast-growing rhizobia express nitrogenase only within the nodule and must therefore be activated by some mechanism that recognizes this situation. It could be under plant control. It has already been pointed out that the electron donors will differ in different organisms. It is to be expected therefore that other organisms will have differences from the genetic system of *Klebsiella*.

7.2.1 *Azotobacter*

Azotobacter has proved very difficult to investigate, as it has 25–40 copies of its chromosome. Thus a single mutation in one copy will not necessarily affect the phenotype. The DNA has been isolated, and by hybridizing with labelled *Klebsiella* DNA from the *nif* region was found to have the *nif* H, D and K genes in a tight cluster. The *nif* M, V and S genes also show homology with *Klebsiella* DNA (Robson *et al.*, 1984).

In spite of the multiple chromosome copies it has been possible to produce some mutant phenotypes. With these, the similarity of the *Klebsiella* genes to those of *Azotobacter* has been examined by inserting the *Klebsiella* genes into a plasmid and introducing the plasmid into an *Azotobacter* mutant to see whether the *Klebsiella* genes will complement the mutation and render the system active again. For example, when a plasmid having the *nif* structural genes is inserted into an *Azotobacter* unable to fix nitrogen, the ability to fix nitrogen is restored. This ability to fix is controlled by fixed nitrogen and thus shows that the regulatory gene products of *Azotobacter* are compatible with the *nif* genes of *Klebsiella* (Cannon and Postgate, 1976; Kennedy and Robson, 1983).

A gene-for-gene comparison of *Azotobacter* and *Klebsiella nif* genes is not appropriate here, but the preceding discussion shows that both structural and control genes have a high degree of homology in these two organisms. This would suggest a high degree of conservation of the gene structures during the course of evolution. The alternative to this is that both organisms have derived the nitrogen-fixing genes from a common source in recent times.

7.2.2 *Rhizobium*

The arrangement of the *nif* genes in different *Rhizobium* species does not at the moment seem to conform to any common pattern. Research is very active on this topic, and as more details emerge a common thread may be revealed or we may come to understand the differences. A number of strains of *Rhizobium* have the *nif* genes and the genes concerned with plant bacterial interaction (the symbiotic genes) sited on a plasmid. These are all fast-growing strains of rhizobia. However, one strain of fast-growing rhizobia which nodulates *Lotus* has been found which appears to have the *nif* genes and the symbiotic genes located on the chromosome. Strains of this organism which were cured of the plasmid were still able to nodulate *Lotus pedunculatus* and fix nitrogen. Plasmids which contain the *nif* genes and the symbiotic genes have not so far been found in any slow-growing rhizobia, so the assumption therefore is that in these bacteria the genes are on the chromosome (Chua *et al.*, 1985).

As well as the differences in the siting of the *nif* genes in the rhizobial strains, there are also differences in the organization of the genes on either the plasmid or the chromosome. Some species such as *R. meliloti* and *R. leguminosarum* have the *nif* H, D and K genes contiguous on a single operon as in *Klebsiella*, while others such as *R. japonicum* have the *nif* H gene separate from the *nif* D and K genes. In this case the separation is by 20 kilobase pairs (Quinto *et al.*, 1985). A gene analogous to *nif* A has been found within this long intervening sequence of *R. japonicum*. For comparison, the *nif* region of *Klebsiella* is 24 kilobase pairs in length.

7.2.3 *Other organisms*

The separation of the *nif* genes is not unusual as it also occurs in the cyanobacteria. In this group, if *Anabaena* is typical, there is a gap between *nif* D and *nif* K. The *nif* H and D genes share the same operon. The separation in this case is by 11 kilobases. This separation is maintained in the vegetative cells, but as the cells mature into heterocysts this intervening stretch of DNA is excised, so that in mature heterocysts which are expressing the *nif* genes the structural genes are in one continuous unit. The excised fragment forms a circle of DNA. It is not known whether any genes are sited on this nor whether, if they are, they are expressed once excised from the *Anabaena* chromosome (Golden *et al.*, 1985).

One further difference in the *nif* gene organization in different organisms is that in some organisms there is more than one copy of one or more of the *nif* genes. *Anabaena* has two copies of *nif* H and *R. phaseoli* has three. There are several copies of the three structural genes in *Rhodopseudomonas capsulata*. Several copies of genes are common in eukaryotes, but multiple copies of genes are relatively rare in prokaryotes, so that the significance of multiple copies cannot as yet be assessed.

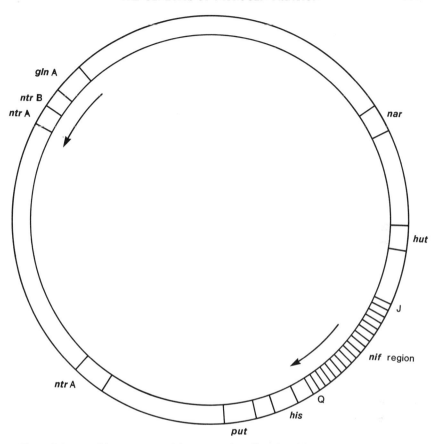

Figure 7.3 Map of the *ntr* genes and those genes that affect the *Klebsiella pneumoniae* genome. Arrows indicate the direction of transcription of the operons. The *his* gene is not controlled by the *ntr* genes but is included as it is an important marker gene.

It is clear that there is a long way to go in the study of the genetics of nitrogen fixation in all these organisms. In the case of *Frankia* the process has hardly begun. The organism grows slowly in culture with a generation time of up to two days. The first aim is therefore to isolate the *nif* genes or *nif* gene region and transfer them to a more quickly-growing organism in order to facilitate their study. So far two plasmids have been isolated from *Frankia* and these have been mapped for the sites at which they are cut by some restriction enzymes. There is, at the moment, no means of introducing plasmid DNA into *Frankia* and it is hoped that modification of one of these two plasmids will enable this to be done. The procedures for the manipulation of DNA in this organism must be worked out before the study of its genetics can be implemented (Normand *et al.*, 1985).

The object of the discussion above has been to show that the *nif* genes have a varied organization and that the system so well worked out for *Klebsiella* is not typical of all. Our knowledge of the *nif* genes in other organisms is on average limited to four of the 17 genes known to be present in *Klebsiella*, the three structural genes *nif* H, D and K and the control gene *nif* A. Clearly the investigation of the genetics of the nitrogen-fixing system in organisms other than *Klebsiella* is yet at an early stage.

7.3 Genes concerned with bacterial–plant interaction

After sowing and inoculating it takes three to four weeks to obtain active nodules on a plant. The testing of strains for their ability to form an effective association is therefore very time-consuming. Earlier effects of mutations on the symbiosis will of course take less time, but the testing of mutants that affect symbiosis is in any event slow. Progress in this field cannot therefore expect to be rapid, and studies have only recently begun. Even so, a large number of investigators is now working in this area, so that many facts are accumulating. The aim of this part of the chapter is not to provide a complete account of the known facts but rather to illustrate some of the research that is going forward and to indicate the usefulness of a genetic approach to an analysis of nodule development.

The nodule structure is well organized and infection and nodule development are complex processes (see Chapter 2). The genetic control of these processes is effected by genes both in the bacterium and in the plant and in some instances the genes may be complementary. An example of this is to be found in the determination of specificity. The bacterial gene determines the specific sugar in its coat and the plant gene specifies the lectin specific for the sugar. Only when these genes match will infection take place. Another example can be deduced from the properties of uptake hydrogenase. The expression of *hup*, the gene, or genes, that codes for uptake hydrogenase, is determined by the host. Thus uptake hydrogenase is expressed in one strain of *R. leguminosarum* when the bacterium is present in pea nodules but not when the bacterium is present in the nodules of broad bean, *Vicia faba* (Dixon, 1972). Hence the correct gene has to be present in the host for *hup* to be expressed in the bacteroid. A consideration of the genetics of the plant and bacterium show that each has genes which affect similar stages in nodule development. Further work will be needed to see how this is brought about and to what extent the genes are complementary in the stages of root infection and nodule development.

7.3.1 Rhizobial symbiotic genes

The symbiotic genes, those concerned with infection and development, are located at the same site as the *nif* genes, that is either on a plasmid or the chromosome. This, however, does not preclude the possibility that when the *nif* genes and the symbiotic genes are present upon a plasmid, relevant genes may also be present upon the chromosome. Indeed there are some indications that this might be so. When a *sym* plasmid from *Rhizobium* is introduced into *Agrobacterium tumefaciens*, it is possible to form legume root nodules. The nodules that are produced by this combination are, however, not effective (Hirsch *et al.*, 1985).

To estimate the number of genes required in the bacterium for nodule formation we need to know in some detail the contribution of both the bacterium and the plant to that process. We do not have enough knowledge at present to hazard a guess. We know that there must be a gene(s) concerned with specificity, gene(s) concerned with hydrogenase, where this is present. Other genes may be control genes affecting different parts of the bacterial metabolism. For example, haem groups are synthesized in the bacterium normally. A different control of the expression of the genes concerned with haem synthesis may be required when haem is synthesized in large amounts for the synthesis of leghaemoglobin.

Some experiments have been done to estimate the amount of DNA required to code for the symbiotic genes and from this to estimate the number of genes required. Pieces of DNA from the *sym* plasmid of *R. leguminosarum* were introduced into a number of plasmids. A strain of *R. leguminosarum* was cured to its *sym* plasmid and the synthesized plasmids which contained parts of the *sym* plasmid introduced into this strain. Two of the plasmids were able to replace the *sym* plasmid, as nodules were formed which contained bacteroids. The amount of DNA that was common to the two plasmids was 10 kilobases. Thus here the genes concerned with nodulation must occupy less than 10 kilobases. For comparison the *nif* region, which contains 17 genes, is 24 kilobases in length (Downie *et al.*, 1984).

Chua *et al.* (1985) inserted a transposable element or transposon, Tn5, into *R. loti*, which nodulates *Lotus pedunculatus*. When a transposon inserts itself into a gene the gene becomes non-functional, as the base sequence is interrupted. By inserting Tn5 into the chromosome 12 mutant strains were obtained of which seven were analysed. One strain was completely defective in nodulation and did not evoke root hair curling. It is probable then that this strain was mutated in the gene responsible for specificity. It showed the *Hac*$^-$ phenotype, where *Hac* is a presumptive hair-curling gene. Five other mutants developed nodules on the *Lotus* but these nodules were aberrant. They contained no *Rhizobium*-infected cells, although rhizobia were found between some of the cortical cells. These were designated as being defective in the *Noi*, nodule initiation, phenotype. Further analysis showed that the Tn5 had entered at least four different sites on the chromosome, so that there is a possibility of four genes being responsible for the *Noi* phenotype. A further mutant formed normal-looking nodules, but on examination of the fine structure it was seen that the bacteria were confined to the infection thread, the mutation preventing release of the bacteria from the infection thread. This phenotype was labelled *Bar*. The remaining three mutants were in the *nif* or *nif*-related genes as normal bacteroids were produced but there was no nitrogen fixation. From this work it would appear that there are at least six genes concerned with nodule development.

7.4 Host symbiotic genes

The long generation time of plants as compared to bacteria has meant that the study of the plant genetics that go hand in hand with those of the bacteria themselves has been slow. Pioneering work by Nutman (1969) showed that there were at least four genes in the genome of *Trifolium pratense* that determined ineffectiveness. These were labelled as *ie*, i_1, *n* and *d*. By crossing plants that contained these genes he was able to show that the ineffective response was recessive in each case and that each of these effects was on a separate locus and hence represented separate genes. These genes were concerned with a mid-stage in nodule development: bacteria had emerged from the infection thread but did not develop into bacteroids. The actual nodule development varied between the mutants. Mutants in the *n* locus were similar to the *Bar* phenotype in that very few cells were infected. The interesting point about these mutations is that expression of the double recessive genotype in the host depended upon the strain of bacterium. With some bacteria, normal effective nodules were produced. Similarly, when the plant was heterozygous for these genes, all the strains produced effective nodules. Thus both the plant and the bacteria had the necessary genes to produce effective nodules, but in some combinations, double recessive plant genes and particular strains of bacteria, only ineffective nodules were produced. Some of the interactions found are shown in Table 7.3.

These interactions suggest that the genes in the bacteria must match genes in the plant for successful nodulation and, depending upon the mismatch, the process of nodulation will break down at different stages.

Table 7.3 Interaction of genes in *Trifolium pratense* and strains of *Rhizobium trifolii*. Homozygotes for each of the host genes, *i, ie, n* and *d*, can form an effective association with at least one strain of bacterium. Each bacterial strain is also capable of forming an effective association. However, some host–strain combinations form ineffective nodules. Data from Nutman (1981).

Rhizobium strain	Heterozygotes	Host genes homozygotes for:			
		i	*ie*	*n*	*d*
47	E	nt	nt	E	E
220	E	E	E	I	I
30	E	E	I	I	I
33	E	I	I	nt	nt

E, effective nodules; I, ineffective nodules; nt, not tested.

Nutman has pointed out, however, that this must not be taken to mean that every gene concerned with nodulation and nitrogen fixation must be matched by the two partners.

Later work on other species has similarly found a number of genes concerned with nodulation and nitrogen fixation. These relate to initial infection and non-nodulating lines of a number of species and genes have been identified that control root hair curling, infection thread development, bacteroid development and the fixation of nitrogen. With these stages in the bacteroid development there are also concomitant changes in the development of the nodule plant cell that are associated with plant genes.

Genes which control the timing of nodulation have also been found. This is of importance agronomically, as if the climate is suitable the early-nodulating combinations will be able to fix nitrogen for a longer time during the season and hence be of more benefit for plant growth.

7.4.1 *Nodulins*

Within the nodule cells proteins are produced that are specific to the nodule and are not found in normal root tissue. These proteins are termed nodulins. The prime example of a nodulin is of course leghaemoglobin. Other nodulins that have been identified are enzymes associated with the metabolism of the fixed nitrogen, but the function of most nodulins is at the moment unknown.

Two methods have been used to identify the number of nodulins in root nodules. One approach, taken by Auger and Verma (1981), was to identify the number of messenger RNAs present in the nodule that were not present in the root. This gives the number of proteins being synthesized at the time the sample was taken. Thus by taking samples at different times during nodule development it was possible to determine the amount of mRNA specific to nodules that were present during the stages of nodule development. In order to find the nodule specific mRNA it was first necessary to make a copy of the total mRNA in the nodule by preparing labelled copy DNA (cDNA). This is done by using the enzyme reverse transcriptase and using labelled nucleotides. Reverse transcriptase synthesizes a complementary copy of DNA from the RNA so that the cDNA has the complementary sequence to the RNA.

The cDNA, which is a copy of the total RNA, is then hybridized to uninfected root RNA. Any cDNA with a complementary sequence to this binds to the mRNA and it is possible to separate out the cDNA that has not bound. This will be the cDNA which has sequences that are not complementary to some of the mRNA in the uninfected root, but they will be complementary to some of the mRNA present in the root nodule and will therefore represent the mRNA for proteins that are specific to the root nodule. Appropriate checks have to be made to ensure that this is so and that the unbound cDNA is not complementary to any rhizobial RNA or DNA. These checks were made and it was determined that between 19 and 40 sequences of cDNA were specific to the nodule and hence that there are between 19 and 40 nodule specific proteins or nodulins. The cDNA sequences will hybridize with the host DNA, showing that these nodulins are all coded for on the host genome. The expression of the genes, the amount of cDNA, increased during nodule development.

The other method, which has similarities in some respects to the first, depends upon the use of antisera to identify nodulins. An antiserum was prepared against the total nodule proteins and then this was reacted with a protein extract of uninfected roots. This removes antisera to the uninfected root proteins and any antisera that remain are specific to root nodule proteins.

An extract of root nodule proteins was then subjected to electrophoresis and the proteins on the gels were transferred by blotting on to nitrocellulose. The transfers were then reacted with the antisera specific to the nodules. These would then bind to the proteins that are specific to the nodule, the nodulins. The antigen–antibody complexes so formed were identified by reacting the complexes with protein A to which ^{125}I was bound to make it radioactive. The radioactivity of the triple complex antibody–antigen–protein A can then be located by overlaying X-ray film. Protein A is a protein obtained from *Staphylococcus aureus* which has the property of binding to rabbit antisera–protein complexes. By using this technique about 30 nodulins were identified, the number and the amount of each differing according to the time of development of the nodule (Bisseling et al., 1984).

It is possible that some of the nodulins may not be really specific to the nodules but may also be present in other parts of the plant at different times. In this case it will be necessary to understand the control mechanisms that enable the genes for these proteins to be processed in the nodule and not in the root.

7.4.2 Leghaemoglobin

Because it is produced in such large quantity in the nodule and because it is related in function to animal haemoglobin and myoglobin, a lot of attention has been given to the genetics of leghaemoglobin. There are four main leghaemoglobin molecules in soya bean root nodules and a few others at much lower concentration. The latter are thought to be derived from the modification of the four main types when once they are synthesized. Four loci have been found which code for these four proteins. These loci are in pairs which are not connected to each other on the scale of the analysis, though it is possible that they are on the same chromosome. It has however been suggested that the four genes are due to the doubling of the chromosomal complement, as soya bean is tetraploid. Thus the genes on the different chromosomes may have mutated slightly differently. The leghaemoglobin molecules only differ in about 10 amino acid residues (Marker et al., 1984).

Of interest from an evolutionary point of view is the fact that the soya bean leghaemoglobin genes have 3 *introns*. The genes of eukaryotes differ from those of prokaryotes in that they have intervening sequences of DNA that do not code for the protein that is coded for by the gene. The mRNA that is synthesized from the gene DNA is thus longer than is required for the protein, and in subsequent processing the intervening sequences are removed from the RNA and the pieces that code for the protein are joined together in order to synthesize the protein for which the gene codes. The parts of DNA that code for the protein are termed exons and the intervening sequences are termed introns. The introns may also code for different proteins, but as space precludes a proper discussion the reader is referred to Danchin and Slonimsky (1985).

Two of these introns are in the same position as the introns in the globin genes of animals. Animal globin genes have two introns. The other interesting point is that the third intron, in the leghaemoglobin gene, is present at the site at which it had been predicted that the animal gene had lost an intron (Go, 1981).

The question of the evolution of the leghaemoglobin gene and the genes for other plant haemoglobins is an interesting one. Haemoglobins are not normally present in plants, and until relatively recently the only known plants to contain this protein were legumes, in which it is found within the root nodule. As these plants are all closely related, the haemoglobin gene could have been acquired at an early stage of evolution of that family. Haemoglobin has now been found to occur in several unrelated families, all of which have nitrogen-fixing symbioses. Analysis of the genes of haemoglobin in these families will help to resolve whether they have acquired the genes separately or whether the genes have entered plants before the families diverged. The root nodule is such a complex and specialized structure that we can only speculate about its evolution.

It will be seen from the discussions in this chapter that the genetics of nitrogen-fixing systems are only just beginning to be understood. For instance, it has now been discovered that substances in plant exudates derepress or 'switch on' genes concerned with nodulation, the *nod* genes (Beringer, 1985). Research on genetics is increasing at a rapid rate as it is realized that hope of improvement of nitrogen-fixing systems for their use in agricultural systems must be based on genetic manipulation of the organisms and associations, whether this be by normal plant breeding methods or by the latest techniques of genetic engineering. The possibilities are discussed in Chapter 8.

CHAPTER EIGHT

FUTURE PROSPECTS

Priority ratings in the research budgets of national and international grant-awarding agencies have facilitated many of the recent advances in the understanding of biological nitrogen fixation which have been discussed in preceding chapters. The resulting framework of data which describes and analyses the nitrogen fixation process is being complemented further by the application to diazotrophic organisms of the newer techniques of genetic analysis, as has been described in Chapter 7. Such studies are providing a better understanding of the ways in which nitrogen fixation is controlled and integrated into cell function and are facilitating exploration of the possibilities for the transfer of the gene systems, which encode the information for nitrogen synthesis and function, into other organisms. It is clear that the availability of such techniques will aid and hasten the integration of new findings concerning, for example, the biochemistry of nitrogen fixation into the emerging functional picture of this process in both free-living and symbiotic systems. The more traditional approaches of applied research in the fields of ecology, agriculture and forestry will, of course, continue to have essential roles in the future, particularly in the vital areas of the translation of laboratory findings to field applications. This chapter will discuss the feasibility of increasing the efficiency of nitrogen-fixing systems that are exploited at the moment, and also the possibilities for widening the range of symbiotic systems that can be used.

8.1 Field applications of nitrogen fixation

The prospects of substantial nitrogen fixation by free-living bacteria held out in the 1970s have not been realized, except in a few special cases, so that symbiotic systems still offer the most effective environment for nitrogen

Table 8.1 Annual increments of nitrogen associated with some free-living associative or symbiotic nitrogen-fixing prokaryotes.

Organism or association	Nitrogen increment (kg/ha)	Reference
Azolla filiculoides	35–63	Talley and Talley (1978)
Cyanobacteria in African soils	3–30	Stewart *et al.* (1979)
Maize/*Spirillum lipoferum*,	2–5	Burris (1977)
Grain legumes (soyabean, bean, cowpea, etc.)	40–354	Franco (1979)
Alnus spp.	40–362	Tarrant (1983)
Casuarina equisetifolia	40–60	Gauthier *et al.* (1985)

fixation. Rates of fixation, compiled for different nitrogen-fixing systems, are compared in Table 8.1.

There are two main approaches by which greater utility of nitrogen fixation could be achieved. First, existing nitrogen-fixing systems might be manipulated to increase the economic returns from crop plants. Second, as yet unutilized or under-utilized systems might be harnessed, or improved for crop production. The first approach covers a wide range of possibilities, important among which are the following.

1. Genetic manipulation of diazotrophic microorganisms and host plants—future prospects for this are considered below. Included under this heading is the selection of host plant and microsymbiont, by which symbiotic partners may be best matched to each other and to particular environments.

2. Novel inoculation procedures. Considerable effort has been devoted to the identification of strains of rhizobia with broad host plant specificity and to improving legume inoculants to ensure survival of rhizobia after introduction into soils. *Rhizobium* inoculation has traditionally involved the coating or pelleting of seeds with rhizobia prior to sowing, or the inoculation of soils with peat-based inoculum or gypsum granules. Another approach which has been developed by Dommergues and co-workers involves the entrapping of rhizobia in polymeric gels such as polyacrylamide, or in much cheaper calcium alginate beads (Figure 8.1). These preparations can be used for soil or seed inoculation and ensure good survival of rhizobia. They have also proved useful as carriers for *Frankia*. Further development of these procedures may help to ensure long-term survival of symbiotic diazotrophs after soil inoculation.

There is now good evidence of interactions between ecto- and endo-

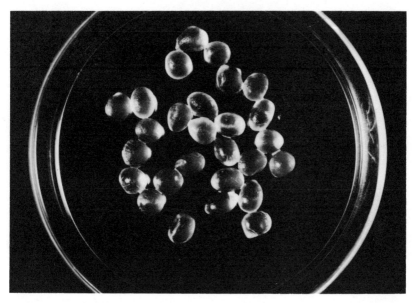

Figure 8.1 Alginate beads, gelled with calcium chloride and incorporating a suspension of *Frankia*.

mycorrhizas and nitrogen-fixing root nodules that are beneficial to host plant growth. The ability of plants to benefit from extra nitrogen depends to a large extent upon the availability of other nutrient elements, and the mycorrhizas contribute largely by increasing the assimilation of these elements, particularly phosphorus, from the soil. Reproducible techniques for the culture of endomycorrhizal organisms have not as yet been established. The availability of such techniques should hasten the development of methods which can be used commercially to inoculate crop or tree species with suitable combinations of both mycorrhizal and nitrogen-fixing organisms. The observations of Knowlton *et al.* (1980) of enhanced nodulation of alders by *Frankia*, due to the induction of root hair deformation when this organism was inoculated together with a soil pseudomonad, suggest that the effectivity of plant inoculation may be improved further by the inclusion of such 'helper' bacteria in the inoculum preparations. The suggestion by Trappe and others that the symbionts of trees such as alders may antagonize the growth of populations of some tree pathogens is also worthy of further consideration. Undoubtedly there are other as yet undiscovered ways in which symbiotic microorganisms and the general soil microflora interact, and advantage might be taken of these to improve plant growth in ways which may not necessarily involve direct effects on mineral nutrition.

8.2 Plant cropping systems

Much consideration is being given at the present time, especially in some of the underdeveloped tropical and subtropical countries, to the further development of multiple cropping or intercropping systems to enhance agricultural production. These techniques are designed to make full use of both the land available for cultivation and of the growing period of the crop.

Although the basic practices of intercropping are often centuries old, it is only recently that they have become the subject of intense interest by the scientific community concerned to maximize crop production. Nitrogen-fixing species clearly play an essential role in the maintenance, by biological means, of the fertility of intensively cropped land as a result of the decay of plant residues and in some cases by excretion of nitrogen-rich root exudates. They are, of course, of major importance in providing nitrogen-rich crops for consumption. Research of a long-term nature is in progress in many parts of the world to obtain the maximum productivity from these systems by defining such factors as the timing of cropping intervals, the most suitable combinations of nitrogen-fixing and non-nitrogen-fixing species for intercropping systems, and the optimum plant spacings within such systems, and also to determine what proportion of the nitrogen fixed by an intercropped nitrogen-fixing species may become available to associated crop plants or to enhance soil fertility following the decay of plant residues.

Woody as well as herbaceous plants may be utilized in multiple cropping systems, and examples of the use of two tree species of particular importance in the economy of many tropical countries are illustrated in Figures 8.2 and 8.3. *Casuarina* is one of the major sources of firewood and light construction timber in the tropics/subtropics and is a tree species of vital importance to millions of people. The photograph shows part of an experiment in progress in Southern India to establish the best techniques for intercropping young *Casuarina equisetifolia* in the traditional seven-year rotation of this species. The second photograph shows a mature plantation of the legume tree *Leucaena leucocephala* which was intercropped during the first years with three pulse or grain cash crops. When canopy closure and root competition prohibited intercropping, the plantation was available for forage, as an additional benefit is the copious

FUTURE PROSPECTS 129

Figure 8.2 *Casuarina equisetifolia* intercropped with cotton in experimental plantings of the Forest Research Station, Tamil Nadu Agricultural University, India.

Figure 8.3 Cattle grazing seedlings beneath the canopy of a *Leucaena leucocephala* plantation, in South India, intercropped prior to canopy closure.

seedling production beneath the canopy, which can be grazed by cattle. The trees themselves will finally be harvested to yield timber useful for poles or firewood.

Research in temperate regions also seeks to further the use of nitrogen-fixing trees in forest ecosystems. Intermixes of nitrogen-fixing and non-nitrogen-fixing trees are being studied to obtain the optimum species combinations and silvicultural conditions. Fast-growing species such as the North American red alder are attractive candidates for short-rotation forestry, which has as its aim the production of wood in dense plantations for pulping or perhaps as fuel. Apart from the ecological desirability of the process, one of the major advantages of the use of nitrogen-fixing trees in such agroforestry systems is the input of a supply of nitrogen over a long period. Against this must be set the slower growth response compared with that due to applied fertilizer nitrogen. This is because of the time taken to render the nitrogen present in the plant residues of the symbionts into a form that is available for the non-nitrogen-fixing species interplanted with them. The ecosystems that result from the mixture of nitrogen-fixing species with others also requires very careful management to obtain the best results. Further detailed consideration of the use of nitrogen-fixing species in multicropping systems and in agroforestry may be found by consulting the publications listed in the references.

8.3 New species

Exploration of the possibilities for exploiting legumes, which up until now have been largely ignored by agriculturalists in developed countries, formed the subject of a special publication, *Tropical legumes: resources for the future* (US National Academy of Sciences, 1979). This report points out that fewer than twenty of the thousands of known legume species are used extensively today and considers a number of root crops, pulses, fruit, forage and timber species to hold particular promise for further research and development. Such reports have stimulated a new interest in the thorough investigation and cataloguing of legumes and other nitrogen-fixing species in countries where the flora is not so thoroughly documented as in most developed countries. Some of the findings to emerge from individual or organised surveys are being catalogued in centralized data banks such as the registry of nodulation of tree species maintained by the University of Hawaii NIFTAL project. The availability and further development of such data systems is likely to be of great benefit to scientists concerned with the range of plant species capable of symbiotic nitrogen fixation.

Literature surveys continue to reveal early reports of nodulation of species rediscovered in recent years and showing symbiotic nitrogen fixation. A survey by Akkermans of the botanical records of the Herbarium Bogorensis, Bogor, Indonesia, revealed the presence of nodulated specimens of *Parasponia* documented and preserved by Clason in the 1930s, some 40 years earlier than the observations of Trinick. Subsequent to communications from Chaudhury of nodulation of *Datisca* in Pakistan, literature surveys by Wheeler in 1978 revealed much earlier reports in the Italian literature of the occurrence of nodules on members of this genus in Europe, which again had been subsequently ignored. Allen and Allen (1958) in their review of nodulated species list a report by Steyaert (1932) of nodules on *Coffea klaini* and *C. robusta* growing in the Congo basin, from which microorganisms with a morphology remarkably similar to *Rhizobium* were isolated. To the authors' knowledge this report has not been pursued further. The possibility that it is due to an *Agrobacterium*-like infection cannot be excluded, but further investigation is merited as *Coffea robusta*, although now planted widely outside Africa, originated in the Congo basin. Another species of importance as a crop plant which is nodulated very locally in Java is *Rubus ellipticus*, the golden evergreen raspberry. Confirmation of nodulation of this species by an organism with the characteristics of *Frankia* has been obtained by Becking (1984), but searches in many other parts of the world have failed to reveal nodules, so that some workers have raised doubts concerning the observation. However, the possibility still remains that in these two cases, and presumably in others, indigenous microsymbionts may be able to form associations with these species, but that elsewhere microsymbionts with the same characteristics may be absent. Diligent inspection of plant root systems may reveal the localized occurrence of nitrogen-fixing nodules on other species, the further study of which might not only help to extend the range of species known to fix nitrogen but may also provide a valuable tool to study host plant–microsymbiont recognition phenomena.

8.4 Genetic engineering

Given the pace of advances in DNA manipulation, the transfer of genes from one prokaryote to another and from prokaryotes to eukaryotes, it would be foolhardy to attempt to predict what will be possible or impossible in a few years' time. We can, however, consider what would be desirable to achieve and what difficulties need to be overcome to attain these objectives. It is the aim of this section to consider the possibilities and to describe, in broad outline, the available methods as we know them today.

A start was made in the transfer of nitrogen-fixing genes into non-nitrogen fixing microorganisms when R.A. Dixon and Postgate (1972) transferred the nitrogen-fixing genes into *Escherichia coli*. They chose a strain of *E. coli* that was deficient in restriction enzymes, so that any DNA transferred would not be cut and rendered inoperative, and that was also his^- (that is, did not synthesize histidine) and was resistant to the antibiotic streptomycin. For the donor strain they chose a strain of *Klebsiella* with an R-factor plasmid, which activates the transfer of chromosomal DNA. As an additional refinement they were able to select an R factor that transferred the *his* locus at high frequency. They had already established that the *nif* genes were closely located to the *his* site on the *Klebsiella* chromosome. The bacteria were mated and then the mixture was plated on to a medium without histidine and which contained streptomycin. Under these conditions only recipient *E. coli* cells that were resistant to streptomycin and that had received the *his* gene from *Klebsiella* could grow. The his^+ strains of *E. coli* were then grown on nitrogen-free medium to select those that had the capability to fix nitrogen. Tests with acetylene and ^{15}N reduction confirmed that these *E. coli* strains had received the *nif* genes from *Klebsiella*.

In later experiments Postgate and his co-workers introduced the *nif* genes into self-transmissible plasmids, which had a wide host range, and were able to transfer the nitrogen-fixing capability by transfer of the plasmids rather than by transfer of chromosomal DNA. Bacteria to which they transferred nitrogen-fixing ability in this way were *Salmonella typhimurium*, *Serratia marcescens* and *Erwinia herbicola*. The *nif* genes in the background of the *Salmonella* genome behave differently in response to the addition of combined nitrogen compounds to the medium. Also, although the plasmid was transferred to *Rhizobium meliloti* and to *Proteus mirabilis*, the *nif* genes were not expressed in either of these two organisms. This shows that the *nif* genes on the plasmid are not sufficient in themselves but have to be controlled by genes on the host's genome. The ability to transfer the *nif* genes by plasmid carriers has led to the main advances that we have in understanding the genetics of the *nif* region, discussed in Chapter 7. It is, however, doubtful whether it is worthwhile to transfer nitrogen-fixing ability to other prokaryotes, as the possibilities for nitrogen fixation by free-living microbes are limited mainly to a few anaerobic sites which have a plentiful supply of energy-rich substrates and for which it is most likely that a suitable natural nitrogen-fixing organism is already available.

8.4.1 Introducing nitrogen-fixing genes into plants

Nitrogen is often the limiting nutrient for plant growth and this has led to the aim, expressed by many writers, of introducing the nitrogen-fixing genes into plants. This aim is laudable but its achievement is not without its problems and these are discussed below.

1. *Physiology*. The effects that the ability to fix nitrogen would have on plant growth and development are not easily predictable and would undoubtedly vary from species to species. The high energy demands of nitrogen fixation could reduce plant growth if the nitrogen-fixing process were not working at maximum efficiency and therefore lowest energy cost. Under these conditions the reduction of nitrogen uses little more energy than the reduction of nitrate (see below), but if the efficiency should be low then the energy demands could be critical at the early stages of plant growth. The energy demands of nitrogen fixation and nitrate reduction are:

$$N_2 + 8e + 8H^+ + 16ATP = 2NH_3 + H_2 + 16ADP + 16Pi$$

with a P:O ratio of 3, 8e is equivalent to 12 ATPs.

Reduction of nitrogen requires 28 ATP equivalents per N_2.

$$NO_3 + 8e + 10H^+ = NH_4^+ + 3H_2O$$

As above, 8e is equivalent to 12 ATPs.

Reduction of nitrate requires 24 ATP equivalents per N_2.

2. *The properties of the nitrogen-fixing system*. The problems posed by the demands of the nitrogen-fixing system for anaerobiosis, low-potential electron donors and a high energy requirement, have been met in various ways by natural nitrogen-fixing organisms. If these genes are introduced into a plant these demands will have to be satisfied.

There are no sites within a plant that are anaerobic and also low-potential electron donors. Chloroplasts and some plastids in roots contain ferredoxins that could supply electrons at a sufficiently low potential. Chloroplasts are, however, very aerobic and the root plastids do not have the respiratory potential to maintain their interior in an anaerobic state.

Mitochondria do have the respiratory activity necessary to maintain their interior in an anaerobic state, but not at all times, although possibility the respiratory demand for energy of the nitrogen-fixing process may maintain respiration rates at the necessary level. Mitochondria will only be active in certain tissues. In order to overcome these problems it would thus be necessary, when engineering a nitrogen-fixing mitochondrial system, to provide controls which would permit the expression of the nitrogen fixation genes in active mitochondria only, otherwise the nitrogenase enzymes synthesized would be inactivated by oxygen in those mitochondria that are unable to maintain a high respiratory rate.

The mitochondria also lack an electron donor of sufficiently low potential to reduce nitrogenase. It would thus be necessary also to provide them with the capacity to synthesize such a donor, together with the enzymes which would facilitate the transfer of electrons to it from a suitable substrate, such as pyruvate. Finally control must be provided to partition mitochondrial metabolism between nitrogen and oxygen reduction. Methods for achieving this may become apparent when the way this is done in organisms such as *Azotobacter* are fully explored.

3. *Gene transfer.* The above discussion shows that, in order to modify the mitochondrion to permit the occurrence of nitrogen fixation, not only do the *nif* genes have to be transferred but also genes for electron transfer from metabolites, together with genes for controlling and integrating the operation of the nitrogen-fixing system into the metabolism of the mitochondrion. With all this to be done, the difficulties associated with the actual transfer of DNA to the mitochondrion and its integration into the mitochondrial genome must seem minor. So far nobody has succeeded in transferring DNA into the mitochondrial genome, but there is every prospect that this will soon be achieved. One problem will be that in a number of cases the coding sequence of the bases in mitochondria is different to that of prokaryotes. Maize mitochondria use the base triplet CGG to code for tryptophan, whereas CGG normally codes for arginine (Leaver and Gray, 1982). This will not matter if these amino acids occur on a non-conserved part of the protein, but if the change alters protein folding or occurs at an active site the protein could be non-functional. The coding sequences for plant mitochondria have not been thoroughly investigated yet and there are suggestions that coding may well differ in the mitochondria of different species of plants. These difficulties are not necessarily insurmountable, as it is now possible to synthesize DNA sequences *in vitro*. The oligonucleotides (short lengths of DNA) that are synthesized can then be introduced into a gene to make specific changes. This technique is discussed below.

4. *Changing genes.* In the past, mutations were produced by chemical or physical means affecting the whole genome. From the mixture of mutations

produced, the geneticist then had to select a desired mutation which, because of the lack of precision in the method, was usually the non-function of a particular gene. We now know the amino acid sequence of a large number of proteins and we also know the base sequence of the genes that code for these proteins. It is thus now possible to insert mutations that are precisely characterized. This technique is known as site-directed mutagenesis.

In order to perform site-directed mutagenesis in this way an oligonucleotide is synthesized which contains the desired mutation. For example, if it is desired to exchange a charged amino acid (glutamic acid) in a protein for a neutral amino acid (alanine), then, in the oligonucleotide, the base sequence GAA for glutamic acid is changed to GCU for alanine. The next step is to insert the new DNA sequence into the gene. This is done by the following steps which are illustrated in Figure 8.4.

1. A site for a restriction enzyme is identified close to the bases that it is desired to change. The DNA is then cut with the restriction enzyme.
2. The two DNA strands are separated by heat treatment and the synthetic oligonucleotide is then annealed to its complementary strand.
3. The rest of the second strand is then synthesized by repairing it with the large sub-unit of *E. coli* DNA polymerase, the Klenow fragment, and rejoined to the original DNA using DNA ligase. A double strand of DNA is thus obtained in which one strand contains the mutation.
4. The DNA is introduced into a bacterium by using a suitable vector, a phage or plasmid. One strand of DNA will not be mutated and will give the wild type and the other will be mutated so that only half, or less of the bacteria will carry the mutant gene, as some of the mutations may be repaired.
5. Colonies of bacteria that contain the mutant gene are identified and grown to supply a quantity of the mutant gene. The means of identifying colonies which contain the mutant gene will differ according to the circumstances. One method is to use labelled oligonucleotide complementary to that previously synthesized, and hybridize this with the bacterial DNA which has been replicated on to nitrocellulose. This will hybridize more efficiently to the mutated genes and thus label them more strongly.
6. The DNA is then introduced into bacteria from which the wild-type gene has been deleted so that the mutant gene can be expressed.

Site-directed mutations are of use in exploring the genetics. Changes can be made in the promoter regions of certain genes to investigate control

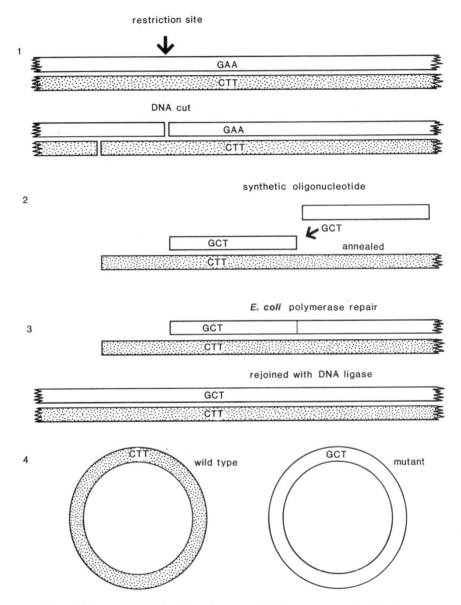

Figure 8.4 One method of site-directed mutagenesis. The steps are outlined in the text.

structures and the cause of the effect produced will be precisely known. Changes can be made to proteins to investigate the effects that certain amino acids may have, either on the protein structure or binding of substrate or ligands, or on activity, etc. The active binding sites of proteins can be explored in this way. Finally, changes can be made to improve the efficiency of enzymes. It may be possible to affect the specificity of proteins by changing the binding sites, for example. Changes that may be desirable for nitrogen fixation could be to make the system less temperature-sensitive; nitrogenase can operate at 37°C but the enzyme cannot be synthesized at this temperature. It may be possible to create enzyme structures that are less sensitive to oxygen and also less sensitive to hydrogen inhibition.

At least at the early stages, genetic engineering will be applied to symbiotic microorganisms as they are the most important in agriculture and forestry. The improved microorganisms have then to infect and nodulate plants. In most cases the microsymbiont will have to compete with indigenous soil microorganisms. In the past, rhizobia have been labelled, by introducing antibiotic-resistance marker genes, and used in field experiments. In many cases the labelled strains have not proved competitive with indigenous rhizobia. In some poor soils not only do the indigenous rhizobia compete successfully with the introduced, superior strains, but may also be completely ineffective. The indigenous rhizobia will have been selected for survival in the particular soil conditions, so that it is not surprising that they will have a competitive edge over an introduced strain. Alternatives to overcome this problem are to inoculate with a high concentration of the strain to be introduced each time the crop is planted, or to engineer both host and microsymbiont with a new specificity that will exclude the indigenous bacteria. In order to obtain the benefits of genetically-engineered improvements it is vital that the bacteria are able to function well in the environment.

8.4.2 *Nodulating new plants*

The introduction and expression of the nitrogen-fixing genes in plants is evidently not a simple undertaking, and it is possible that the problems involved in extending the host range of effective nodulation of *Rhizobium* or *Frankia* might be solved more readily. The principal advantage of this approach is that the root nodule already provides an environment in which the requirements of the nitrogen-fixing system are satisfied.

The difficulty in extending nodulating ability for example to cabbages with *Rhizobium* or to pines with *Frankia*, is that we do not know what plant genes to transfer. It might be relatively straightforward to find the genes which code for the nodulins by sequencing the proteins and then deducing the DNA code, synthesizing the DNA and finding homology with host plant DNA. The genes that are concerned with nodule structure will be less amenable to detection. If all the genes that code for nodulation properties are linked together in the plant genome these problems might be simplified.

Even if it were possible to nodulate other plants, however, it might not always be desirable for plant growth to extend the *Rhizobium* association. For example, in plants which currently have a nitrogen-fixing symbiosis, some time elapses between initial infection and the first fixation of nitrogen. In the interim, plants will require nitrogen and also the resources to develop nodules. Legumes are able to overcome this period of nitrogen shortage in early establishment by drawing on the high nitrogen reserves of the seed. Cereals, however, have a relatively low nitrogen content and require nitrogen at an early stage of growth. If nitrogen is not supplied at an early stage, in the first week, then tiller growth is not supported and the yield of grain decreases considerably (Dale, 1979). No doubt one could find similar problems with a number of plant species that have evolved differently from legumes.

The preceding discussion is not intended to discount the prospect of genetic engineering as a tool to increase nitrogen input into crop plants by nitrogen fixation, but rather to illustrate the extent of the problems to be solved before this is achieved. Meanwhile, research with the more modest aim of improving the nitrogen-fixing systems that we have, should yield valuable results. We can illustrate this approach by considering a recent proposal of Dandekar *et al.* (1983) for a technique to increase nitrogen fixation in legume root nodules under water stress. In 1983 about 33 per cent of the soya bean crop was lost through drought, and it has been estimated that, in any year, dry spells of short duration will decrease yields. If nitrogen fixation could be improved when the nodules are under water stress then the total amount of nitrogen fixed over a season should be increased, with a concomitant improvement in yield.

A number of plants increase their proline content when under water stress, and some salt-tolerant plants (halophytes) normally have a high proline content when growing under conditions of high salt. It is believed that proline acts by decreasing the cell's water potential to that of the surroundings, so that more water may be taken up. If root nodules could be stimulated to produce proline as the result of nitrogen fixation when under water stress, then drought resistance of the plants may well be improved.

Experiments supporting this view have been done in which nodules, water-stressed by putting salt in the growth medium, were partially protected by the addition of glycine betaine or proline betaine to the growing medium. These two compounds are also thought to be taken up by the nodulated root system where they act like proline.

The ability to induce root nodule microsymbionts to excrete proline has been made possible by the discovery of a mutant of *Escherichia coli* which overproduces proline. Feedback inhibition in the synthesis pathway is considerably lessened in this mutant. When this mutant gene was put into a plasmid and transferred to *Klebsiella* it was found that possession of the proline over-producing genes relieved water stress in this organism. Some of the experimental results are given in Table 8.2. Analysis of the amino acids in the strains with and without the mutant gene showed that they contained 429 and 42 nmoles proline per mg protein respectively. This result, together with the experiments in which water stress was relieved in whole nodules by uptake of betaines from the growth medium, gives good grounds for hope that insertion of the proline over-producing gene into the *Rhizobium* genome will help to protect the nodules under stress. However, the nodule bacteroids would not normally be able to synthesize proline,: the nitrogen that is fixed is excreted into the host cell as ammonia as the bacteroids have a very limited means of assimilating ammonia for themselves. To enable the bacteroids to synthesize proline, it would be necessary for the plasmid that carries the proline over-producing gene also to carry genes for the assimilation of ammonia into glutamate. Expression of these genes would have to be controlled to allow excretion of a substantial proportion of the fixed nitrogen for assimilation in the host cell cytoplasm.

A start has been made on techniques of engineering to achieve these aims. The genes have been put into *Klebsiella* and have produced the desired

Table 8.2 The effect of osmotic stress on nitrogen fixation by *Klebsiella*.

	Nitrogenase activity (μmole C_2H_4 hr^{-1} mg protein^{-1})		
NaCl (molar)	KY1 ($-$)	KY2 ($-$)	KY2 ($+$)
0.0	2.64	2.71	2.59
0.3	1.55	0.60	1.31
0.4	1.53	0.14	0.74

($+$), ($-$) With and without exogenous proline
KY1, strain with proline over-producer plasmid
KY2, control strain.

effect. They have now been transferred to some strains of *Rhizobium*, and work is in hand to find whether the genes are expressed in the bacteroids and whether in that case there is benefit to the plant.

The genetic engineering of bacteria has reached the stage where it is almost a routine practice, but the genetic engineering of eukaryotes is just beginning to be explored. As one might expect, there are fresh problems to be solved and these lie in three main areas:

1. The identification and detection of genes. As has already been discussed, the genetics of the plant interaction are poorly understood at present and we need to know which genes to transfer and how to identify them in the plant genome.
2. The introduction and integration of genes into the plant. Some methods of introducing new genes into plants are outlined below.
3. The control of gene expression.

1. *Plasmid transfer.* *Agrobacterium tumefaciens*, the crown-gall-forming organism, contains a plasmid, the Ti plasmid. When it infects a plant, part of this plasmid, T DNA, is transferred to the plant cells and becomes incorporated into the plant chromosomal DNA. The mechanism by which this is done is not yet understood. There are several genes present in the T DNA: these include the *onc* genes and genes that code for particular amino acids, the opines (octopine or nopaline). The *onc* genes enable crown-gall cells to grow without added plant growth regulators, auxin and cytokinin. There is now evidence that the *onc* genes code for the enzymes that are involved in the synthesis of these plant growth regulators and that it is the imbalance of the growth regulators that leads to the proliferation of the cells, thus resulting in tumour formation.

It is possible to remove the Ti plasmid from *Agrobacterium* and insert DNA segments (genes) into the T DNA region of the plasmid. The altered plasmid can then be restored to the bacterium, which can be used to infect a plant. The new T DNA region will then be incorporated into the plant nuclear DNA. If it is desired, most of the T DNA region of the plasmid can be removed to make room for the new genes that are to be inserted (see Figure 8.5). It is often better to leave the genes concerned with opine synthesis, as the presence of octopine or nopaline in the plant tissue shows that plasmid transfer has taken place: in other words, it acts as a marker. If the genes for opine synthesis are removed, some other marker gene may be added. When the genes are inserted into the plasmid they need to be flanked with regulatory sequences of DNA so that they can be transcribed, and the

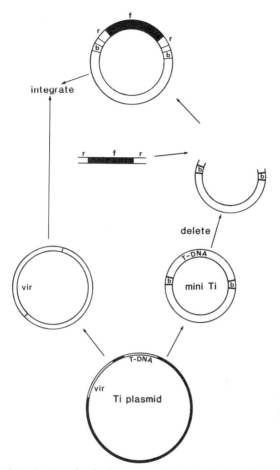

Figure 8.5 The introduction of a foreign gene, f, into the Ti plasmid of *Agrobacterium tumefaciens*. The virulence region and the T-DNA are separated on to two small plasmids. The genes from the T-DNA region are then removed and replaced by the foreign gene. The foreign gene is flanked by two regulatory sequences, r, which enable the gene to be expressed within the plant. The virulence plasmid contains genes that are needed for the insertion of the T-DNA, and foreign gene, into the plant. Reproduced with permission from Grierson and Covey (1984).

gene expressed within the plant. It has been shown that genes added to a plant in this way are carried through meiosis and are present in subsequent generations.

Until recently it was thought that *Agrobacterium* infected only dicotyledonous plants as it produced tumours on these plants only and not on monocotyledons. However, it has now been shown that some monocotyledons are capable of infection by *Agrobacterium*, but that tumours are not produced. Infection and gene transfer was demonstrated when plants that were inoculated with virulent *Agrobacterium* were shown to contain octopine or nopaline in their tissues. The particular opine depended upon the strain of *Agrobacterium* and plasmid with which it was infected. Many important crop plants are monocotyledonous, so that if the Ti plasmid becomes an important vector for transferring foreign genes into plants, this finding is highly significant.

2. *Virus transfer and transfer into protoplasts.* The DNA of some plant viruses can be modified to contain extra genes, although in this case the amount of extra DNA that can be incorporated is limited because of the need to encapsulate it within the viral protein coat. When the plant is infected, the virus spreads to almost every cell in the plant. However, genes transferred in this way cannot be inherited, so that infected plants with the new genes would have to be propagated asexually. DNA can be introduced into naked protoplasts if the protoplasts are incubated with the DNA under the correct conditions. This DNA can become incorporated into the protoplast chromosomal DNA and is thus heritable. This method overcomes the disadvantage of the plasmid transfer method, namely the restricted host range of *Agrobacterium*. A problem with both plasmid transfer and direct transfer into protoplasts is that of the regeneration of whole plants from callus cells. This can be achieved in some instances, but so far it has not proved possible for a number of plant species to regenerate whole plants from callus tissue.

3. *Control.* Plant genes are under developmental control. The processes of differentiation and the differences in plant metabolism in different plant organs show that plant genes are only expressed at certain times. It would be necessary to place any genes that might be transferred under control so that they were expressed only at the right time and in the right place. At the moment our ignorance of this type of control is almost complete.

8.5 Industrial nitrogen fixation

A few words should be devoted to the future prospects of industrial nitrogen fixation, as the success of efforts in this field to produce nitrogenous fertilizer efficiently and cheaply will affect the importance and role of biological nitrogen fixation in agriculture. The ability of the natural

Table 8.3 Metal–ligand complexes that can bind and reduce N_2 (Armstrong, 1985)

1.	$[Ru(NH_3)_5(N_2)^{++}]$	First dinitrogen metal complex (Allen and Senoff, 1965)
2.	$[Mo(N_2)_2(PR_3)_4]$	First complex to produce ammonia (Chatt et al., 1975)
3.	$[W(N_2)_2(Ph_2PCH_2PPh_2)_2]$	First complex to produce ammonia catalytically (Pickett and Talarmin, 1985)

R = an alkyl or aryl group, Ph = phenyl.

catalyst, nitrogenase, to reduce nitrogen at ambient pressure and temperature has led to attempts to emulate this with synthetic catalysts based upon molybdenum and related elements such as tungsten and vanadium. A metal–ligand complex which could bind dinitrogen was first found in 1965 (Allen and Senoff, 1965) but the nitrogen on this complex was not reduced. After further progress a complex was derived which could reduce dinitrogen to ammonia, but not in a cyclic manner. The cyclic catalytic reduction of ammonia has now been achieved, so that progress in this work is now evident (see Table 8.3). However, the complexes that are now required for the reduction of nitrogen would be expensive, and economical methods must yet be found. A study of the chemical methods that can be used at ambient temperature and pressure, using molybdenum or tungsten, may well increase our understanding of the function and action of molybdenum in nitrogenase. Attention has also been paid to the nature of the active centre of nitrogenase, FeMoco, and models have been made of various conformations of this and also with ligands other than sulphur.

A large part of the cost of synthesizing ammonia is in the production of hydrogen. It could be possible to link up the biological production of hydrogen to the catalytic reduction of nitrogen to ammonia. Much research work is being done on the production of hydrogen from algae or chloroplasts in reactions dependent upon photosynthesis. If production of hydrogen can be made to function economically with the use of a suitable catalyst and with sunlight as an energy source, then ammonia can be produced closer to the site of use and without the heavy capital cost of the Haber–Bosch process and without the use of non-renewable energy sources. There is, however, still much work to be done before we can reach this happy state of affairs (Gisby and Hall, 1980; Coucouvanis, 1984).

REFERENCES AND FURTHER READING

Chapter 1

Akkermans, A.D.L., Abulkadir, S. and Trinick, M.J. (1978) Nitrogen-fixing root nodules in Ulmaceae. *Nature (London)* **274**, 190.

Callaham, D., Del Tredici, P. and Torrey, J.G. (1978) Isolation and cultivation 'in vitro' of the actinomycete causing root nodulation in *Comptonia*. *Science* **199**, 899–902.

Dalton, H. and Postgate, J.R. (1968) Effect of oxygen on the growth of *Azotobacter chroococcum* in batch and continuous cultures *J. gen. Microbiol.* **54**, 463–473.

Dixon, R.O.D. (1967) The origin of the membrane envelope surrounding the bacteria and bacteroids and the presence of glycogen in clover root nodules. *Arch. Mikrobiol.* **56**, 156–166.

Dobereiner, J., Day, J.M. and Dart, P.J. (1972) Nitrogenase activity and oxygen sensitivity of the *Paspalum notatum–Azotobacter paspali* association. *J. gen. Microbiol.* **71**, 103–116.

Hill, S. (1976) Influence of atmospheric oxygen concentration on acetylene reduction and efficiency of nitrogen fixation in intact *Klebsiella pneumoniae*. *J. gen. Microbiol.* **93**, 335–345.

Hill, S. Turner, G.L. and Bergersen, F.J. (1984) Synthesis and activity of nitrogenase in *Klebsiella pneumoniae* exposed to low concentrations of oxygen. *J. gen. Microbiol.* **130**, 1061–1067.

Hyndman, L.A. and Burris, R.H. (1953) Properties of hydrogenase from *Azotobacter vinelandii*. *J. Bacteriol.* **65**, 522–531.

Lechevalier, M.P. (1984) The taxonomy of the genus *Frankia*. *Plant & Soil* **78**, 1–6.

Moffet, M.L., and Colwell, R.R. (1968) Adansonian analysis of the Rhizobiaceae. *J. gen. Microbiol.* **51**, 245–266.

Nichols, D.E. Williamson, P.C. and Waggoner, D.R. (1980) Assessment of alternatives to present day ammonia technology with emphasis on coal gasification. In *Nitrogen Fixation I* (eds. W.H. Orme-Johnson and W.E. Newton), University Park Press, Baltimore, 43–60.

Postgate, J.R. (1984) New kingdom for nitrogen fixation. *Nature (London)* **312**, 194.

Robson, R.L. (1984) Characterisation of an oxygen-stable nitrogenase complex isolated from *Azotobacter chroococcum*. *Biochem. J.* **181**, 569–575.

Subba Rao, N.S. (1980) Crop responses to microbial inoculation. In *Recent Advances in Nitrogen Fixation* (ed. N.S. Subba Rao), Edward Arnold, London, 406–420.

Tjepkema, J.D. Ormerod, W. and Torrey, J.G. (1981) Factors affecting vesicle formation and acetylene reduction (nitrogenase activity) in *Frankia* sp. Cpl1. *Can. J. Microbiol.* **27**, 815–823.

Woese, C.R. (1981) Archaebacteria. *Scientific American* **244**, 94–106.

REFERENCES AND FURTHER READING

Chapter 2

Angulo, A.F., Van Dijk C. and Quispel, A. (1976) Symbiotic interactions in non-leguminous root nodules. In *Symbiotic Nitrogen Fixation in Plants* (ed. P.S. Nutman), IBP. 7, Cambridge University Press, Cambridge, 475–484.

Appleby, C.A. (1969) Electron-transport systems of *Rhizobium japonicum*. 1. Hemoprotein P-450, other CO reactive pigments, cytochromes and oxidases in bacteroids from N_2 fixing nodules. *Biochim. Biophys. Acta* **172**, 71–78.

Bohlool, B.B. and Schmidt, E.L. (1974) Lectins: a possible basis for the specificity in the *Rhizobium*–legume root nodule symbiosis. *Science* **185**, 269–271.

Bond, G. (1983) Taxonomy and distribution of non-legume nitrogen-fixing systems. In *Biological Nitrogen Fixation in Forest Ecosystems: Foundations and Applications*, (eds. J.C. Gordon and C.T. Wheeler), Martinus Nijhoff, The Hague, 55–88.

Callaham, D. and Torrey, J.G. (1977) Prenodule formation and primary nodule development in roots of *Comptonia* (Myricaceae). *Can. J. Bot.* **55**, 2380–2381.

Callaham, D.A. and Torrey, J.G. (1981) The structural basis for infection of root hairs of *Trifolium repens* by *Rhizobium*. *Can. J. Bot.* **59**, 1647–1664.

Chaboud, A. and Lalonde, M. (1983) Lectin binding on surfaces of *Frankia* strains. *Can. J. Bot.* **61**, 2889–2897.

Dazzo, F.B. and Brill, W.J. (1978) Regulation by fixed nitrogen of host–symbiont recognition in the *Rhizobium*–clover symbiosis. *Plant Physiol.* **62**, 18–21.

Dommergues, Y.R., Diem, H.G., Gauthier D.L., Dreyfus, B.L. and Cornet, F. (1984) Nitrogen-fixing trees in the tropics: potentialities and limitations. In *Advances in Nitrogen Fixation Research* (eds. C. Veeger and W.E. Newton), Martinus Nijhoff, The Hague, 7–14.

Gordon, J.C. (1984) Biological nitrogen fixation in temperate zone forestry: current use and future potential. *Ibid.*, 15–22.

Grove, T.S., O'Connell, A.M. and Malajczuk, N. (1980) Effects of fire on the growth, nutrient content and rate of nitrogen fixation of the cycad *Macrozamia riedlii*. *Aust. J. Bot.* **28**, 271–281.

Hamblin, J. and Kent, S.P. (1973) Possible role of phytohaemagglutinin in *Phaseolus vulgaris*. *Nature, New Biology* **245**, 28–30.

Knowlton S., Berry, A. and Torrey, J.G. (1980) Evidence that associated soil bacteria may influence root hair infection of actinorhizal plants by *Frankia*. *Can. J. Microbiol.* **26**, 971–977.

Lalonde, M. (1979) Immunological and ultrastructural demonstration of nodulation of the European *Alnus glutinosa* (L.) Gaertn. host plant by an actinomycetal isolate from the North American *Comptonia peregrina* (L.) Coult. root nodule. *Bot. Gaz.* **140**, (Suppl.) S35–S43.

Lalonde, M. and Knowles, R. (1975) Ultrastructure, composition and biosynthesis of the encapsulation material surrounding the endophyte in *Alnus crispa* var. *mollis* root nodules. *Can. J. Bot.* **53**, 1951–1971.

Lancelle, S.A. and Torrey, J.G. (1984) Early development of *Rhizobium*-induced root nodules of *Parasponia rigida*. 1. Infection and early nodule initiation. *Protoplasma* **123**, 26–37.

Libbenga, K.R. and Haaker, P.A.A. (1973) Initial proliferation of cortical cells in the formation of root nodules of *Pisum sativum* L. *Planta* **114**, 17–28.

Miller, I and Baker, D. (1985) The initiation, development and structure of root nodules in *Elaegnus augustifolia* L. (Elaeagnaceae). *Protoplasma* **128**, 109–119.

Mitchell, J.P. (1965) The DNA content of nuclei in pea root nodules. *Ann. Bot. N.S.* **29**, 371–376.

Nathalinez, C.P. and Staff, I.A. (1975) On the occurrence of intracellular blue-green algae in cortical cells of the apogeotropic roots of *Macrozamia communis* L. Johnson. *Ann. Bot. NS* **39**, 363–368.

Pate, J.S., Gunning, B.E.S. and Briarty, L.G. (1969) Ultrastructure and functioning of the transport system of the leguminous nodule, *Planta* **85**, 11–34.

Peters, G.A., Kaplan, D. and Calvert, H.E. (1984) Solar powered N_2 fixation in ferns: the *Azolla–Anabaena* symbioses. *Trans. Roy. Soc. Edin.* **86** (B), 169–178.

Price, G.D., Mohapatra, S.S. and Gresshof, P.M. (1984) Structure of nodules formed by *Rhizobium* strain ANU289 in the nonlegume *Parasponia* and the legume siratro (*Macroptilium atropurpureum*). *Bot. Gaz.* **145**, 444–451.

Robson, R.L. (1979) Characterisation of an oxygen-stable nitrogenase complex isolated from *Azotobacter chrococcum*. *Biochem. J.* **181**, 569–575.

Silvester, W.B. (1976) Endophyte adaptation in *Gunnera–Nostoc* symbiosis. In *Symbiotic Nitrogen Fixation* (ed. P.S. Nutman), Cambridge University Press, Cambridge, 521–538.

Stewart, W.D.P. and Rowell, P. (1977) Modifications of nitrogen-fixing algae in lichen symbioses. *Nature* **265**, 371–372.

Torrey, J.G. and Callaham, D. (1978) Determinate development of nodule roots in actinomycete-induced root nodules of *Myrica gale*. *Can. J. Bot.* **56**, 1357–1364.

Vandenbosch, K.A. and Torrey, J.G. (1985) Development of endophytic *Frankia* sporangia in field- and laboratory-grown nodules of *Comptonia peregrina* and *Myrica gale*. *Amer. J. Bot.* **72**, 99–108.

van Dijk, C. (1979) Spore formation and endophyte diversity in root nodules of *Alnus glutinosa*. *New Phytol.* **92**, 215–220.

Further Reading

Akkermans, A.D.L. and van Dijk, C. (1981) Non-leguminous root-nodule symbioses with actinomycetes and *Rhizobium*. In *Nitrogen Fixation I. Ecology* (ed. W.J. Broughton), Clarendon Press, Oxford, 57–103.

Allen, O.N. and Allen, E.K. (1981) *The Leguminosae: a sourcebook of characteristics, uses and nodulation*. University of Wisconsin Press. Madison.

Millbank, J.W. (1984) Nitrogen fixation in lichens. In *Current Developments in Biological Fixation* (ed. N.S. Subba Rao), Edward Arnold, London, 197–219.

Watanabe, I. and Roger, P.A. (1984) *Ibid.*, 237–276.

Chapter 3

Carter, K.R., Rawlings, J., Orme-Johnson, W.H., Becker, R.R. and Evans, H.J. (1980) Purification and characterization of a ferrodoxin from *Rhizobium japonicum* bacteroids. *J. Biol. Chem.* **255**, 4213–4233.

Dilworth, M.J., (1966) Acetylene reduction by nitrogen-fixing preparations from *Clostridium pasteurianum*. *Biochim. biophys. Acta* **127**, 285–294.

Emerich, D.W. and Burris, R.H. (1978) Complementary functioning of the component proteins of nitrogenase from several bacteria. *J. Bacteriol.* **134**, 936–943.

Haaker, H. and Veeger, C. (1977) Involvement of the cytoplasmic membrane in nitrogen fixation by *Azotobacter vinelandii*. *Eur. J. Biochem.* **77**, 1–10.

Lowe, D.J. and Thorneley, R.N.F. (1984) The mechanism of *Klebsiella pneumoniae* nitrogenase action. *Biochem. J.* **244**, 877–886.

McNary, J.E. and Burris, R.H. (1962) Energy requirements for nitrogen fixation by cell free preparations from *Clostridium pasteurianum*. *J. Bacteriol.* **84**, 598–599.

Nichols, D.E., Williamson, P.C. and Waggoner, D.R. (1980) Assessment of alternatives to present-day ammonia technology with emphasis on coal gasification. In *Nitrogen Fixation* 1 (eds. W.E. Newton and W.H. Orme-Johnson), University Park Press, Baltimore, 43–60.

Robson, L.R. and Postgate, J.R. (1980) Oxygen and hydrogen in biological nitrogen fixation. *Ann. Rev. Microbiol.* **34**, 183–208.

Schubert, K.R. and Evans, H.J. (1976) A major factor affecting the efficiency of nitrogen fixation in nodulated symbionts. *Proc. nat. Acad. Sci. USA* **73**, 1207–1211.

Schollhorn, R. and Burris, R.H. (1967) Acetylene as a competitive inhibitor of N_2 fixation. *Proc. nat. Acad. Sci. USA* **57**, 1317–1323.
Shah, V.K., Chisnell, J.R. and Brill, W.H. (1978) Acetylene reduction by the iron-molybdenum cofactor from nitrogenase. *Biochem. Biophys. Res. Commun.* **81**, 232–236.
Zumft, W.G., Mortensen, L.E. and Palmer, G. (1974) Electron-paramagnetic-resonance studies on nitrogenase. *Eur. J. Biochem.* **46**, 525–535.

Chapter 4

Appleby, C.A. (1984) Leghaemoglobin and *Rhizobium* respiration. *Ann. Rev. Plant Physiol.* **35**, 443–478.
Atkins, C.A. (1974) Occurrence and some properties of carbonic anhydrase from legume root nodules. *Phytochemistry* **13**, 93–98.
Bergersen, F.J. (1963) The relationship between hydrogen evolution, hydrogen exchange, nitrogen and applied oxygen tension. *Aust. J. Biol. Sci.* **16**, 669–680.
Bergersen, F.J. (1982) *Root Nodules of Legumes: Structure and Functions*. Research Studies Press, Chichester.
Bergersen, F.J. and Goodchild, D.J. (1973) Aeration pathways in soybean root nodules. *Aust. J. Biol. Sci.* **26**, 729–740.
Bergersen, F.J. and Appleby, C.A. (1981) Leghaemoglobin within bacteroid-enclosing membrane envelopes from soybean root nodules. *Planta* **152**, 534–543.
Dilworth, M.J. (1969) The plant as the genetic determinant of leghaemoglobin production in the legume nodule. *Biochim biophys. Acta* **184**, 432–441.
Schubert, K.R. and Evans, H.J. (1976) Hydrogen evolution: a major factor affecting the efficiency of nitrogen fixation in nodulated symbionts. *Proc. nat. Acad. Sci. USA.* **73**, 1207–1211.
Sinclair, T.R. and Goudriaan, J. (1981) Physical and morphological constraints and transport in nodules. *Plant Physiol.* **67**, 143–145.
Tjepkema, J.D. and Yocum, C.S. (1974) Measurement of oxygen partial pressure within soybean nodules by oxygen microelectrodes. *Planta* **119**, 351–360.
Wittenberg, J.B., Appleby, C.A. and Wittenberg, B.A. (1972) The kinetics of the reactions of leghaemoglobin with oxygen and carbon monoxide. *J. Biol. Chem.* **247**, 527–531.

Chapter 5

Akkermans, A.D.L., Roelofsen, W., Blom, J., Huss-Danell, K. and Harkink, R. (1983) Utilisation of carbon and nitrogen compounds by *Frankia* in synthetic media and in root nodules of *Alnus glutinosa, Hippophae rhamnoides* and *Datisca cannabina. Can. J. Bot.* **61**, 2793–2800.
Appleby, C.A., Tjepkema, J.D. and Trinick, M.J. (1983) Haemoglobin in a non-legume plant, *Parasponia*: possible genetic origin and function in nitrogen fixation. *Science* **220**, 951–953.
Atkins, C.A. (1974) Occurrence and some properties of carbonic anhydrase from legume root nodules. *Phytochemistry* **13**, 93–98.
Benson, D.R., Arp, D.J. and Burris, R.H. (1980) Hydrogenase in actinorhizal root nodules and root nodule homogenates. *J. Bacteriol.* **142**, 138–144.
Berg, R.H., (1983) Preliminary evidence for the involvement of suberization in infection of *Casuarina. Can. J. Bot.* **61**, 2910–2918.
Davenport, H.E. (1960) Haemoglobin in the root nodules of *Casuarina cunninghamiana. Nature (London)* **186**, 653–654.
Lechevalier, M.P., Horriére, F. and Lechevalier, H. (1982) The biology of *Frankia* and related organisms. In *Developments in Industrial Microbiology*, Society for Industrial Microbiology, **23**, 51–60.
Lopez, M.F. and Torrey, J.G. (1985) Enzymes of glucose metabolism in *Frankia* sp. *J. Bacteriol.* **162**, 110–116.

McClure, R.P., Coker, G.T. and Schubert, K.R. (1983) Carbon dioxide fixation in roots and nodules of *Alnus glutinosa*. *Pl. Physiol.* **71**, 652–657.
Murray, M.A., Fontaine, S. and Torrey, J.G. (1984) Oxygen protection of nitrogenase in *Frankia* sp. HFP Arl3. *Arch. Microbiol.* **139**, 162–166.
Norman, P. and Lalonde, M. (1982) Evaluation of *Frankia* strains isolated from provenances of two *Alnus* species. *Can. J. Microbiol.* **28**, 1133–1142.
Roelofsen, W. and Akkermans, A.D.L. (1979) Uptake and evolution of H_2 and reduction of C_2H_2 by root nodules and nodule homogenates of *Alnus glutinosa*. *Plant Soil* **52**, 571–578.
Tjepkema, J.D. (1983) Haemoglobins in the nitrogen-fixing root nodules of actinorhizal plants. *Can. J. Bot.* **61**, 2924–2929.
Tjepkema, J.D., Orme, W. and Torrey, J.G. (1981) Factors affecting vesicle formation and acetylene reduction (nitrogenase activity) in *Frankia* sp. Cpl1. *Can. J. Microbiol.* **27**, 815–823.
Vandenbosch, K.A. and Torrey, J.G. (1984) Consequences of sporangial development for nodule function in root nodules of *Comptonia peregrina* and *Myrica gale*. *Pl. Physiol.* **76**, 556–560.
Wheeler, C.T., Ching, T.M. and Gordon, J.C. (1979) Oxygen relations of the root nodules of *Alnus rubra* Bong. *New Phytol.* **82**, 449–457.
Wheeler, C.T., Watts, S.H. and Hillman, J.R. (1983) Changes in carbohydrate and nitrogenous compounds in the root nodules of *Alnus glutinosa*. *New Phytol.* **95**, 209–218.
Winship, L.J. and Tjepkema, J.D. (1983) The role of diffusion in oxygen protection of nitrogenase in nodules of *Alnus rubra*. *Can. J. Bot.* **61**, 2930–2936.

Chapter 6

Boland, M.J. and Schubert, K.R. (1983) Biosynthesis of purines by a proplastid fraction from soybean nodules. *Arch. Biochem. Biophys.* **220**, 179–187.
Boland, M.J., Hanks, J.F., Reynolds, P.H.S., Blevins, D.G., Tolbert, N.E. and Schubert, K.R. (1982) Subcellular organisation of ureide biogenesis from glycolytic intermediates and ammonium in nitrogen-fixing soybean nodules. *Planta* **155**, 45–51.
Dixon, R.O.D. and Wheeler, C.T. (1983) Biochemical, physiological and environmental aspects of symbiotic nitrogen fixation. In *Biological Nitrogen Fixation in Forest Ecosystems: Foundations and Applications*, Martinus Nijhoff, The Hague, 107–171.
Dreyfus, B.L. and Dommergues, Y.R. (1980) Non-inhibition of nitrogen fixation by combined nitrogen is stem bearing nodules in the legume *Sesbiana rostrata*. *C.R. Acad. Sci. D.* **291**, 767–770.
Eady, A., Kahn, D. and Hawkins, M. (1981) Metabolic control of *Klebsiella pneumoniae* mRNA degradation: significance to *nif* gene regulation by NH_4^+. In *Current Perspectives in Nitrogen Fixation* (eds. A.H. Gibson and W.E. Newton), C.S.I.R.O., Canberra, 443.
Halliday, J. and Pate, J.S. (1976) Symbiotic nitrogen fixation by coralloid roots of the cycad *Macrozamia riedlii*: physiological characteristics and ecological significance. *Aust. J. Pl. Physiol.* **3**, 349–358.
Hom, S.S., Hennecke, H. and Shanmugam, K.T. (1980) Regulation of nitrogenase biosynthesis in *Klebsiella pneumoniae* : Effect of nitrate. *J.gen. Microbiol.* **117**, 169–179.
Huss-Danell, K., Sellstedt, A., Flower-Ellis, A. and Sjostrom, M. (1982) Ammonium effects on function and structure of nitrogen-fixing root nodules of *Alnus incana* (L.) Moench. *Planta* **156**, 332–340.
Ingestaad, T. (1980) Growth, nutrition and nitrogen fixation in grey alder at varied rates of nitrogen addition. *Physiol. Plant.* **50**, 353–364.
Laane, C., Krone, W., Konings, W.N., Haaker, H. and Veeger, C. (1980) Short-term effects of ammonium chloride on nitrogen fixation by *Azotobacter vinelandii* and by bacteroids of *Rhizobium leguminosarum*. *Eur. J. Biochem.* **103**, 39–46.

Leaf, G., Gardner, I.C. and Bond, G. (1958) Observations on the composition and metabolism of the nitrogen-fixing root nodules of alder. *J. exp. Bot.* **9**, 320–331.

Martin, F., Hirel, B. and Gadal, P. (1983) Purification and properties of ornithine carbamyl transferase from *Alnus glutinosa* root nodules. *Z. Pflanzenphysiol.* **111**, 413–422.

McClure, P.R. and Israel, D.W. (1979) Transport of nitrogen in the xylem of soybean plants. *Pl. Physiol.* **64**, 411–466.

McClure, P.R., Coker, G.T. and Schubert, K.R. (1983) Carbon dioxide fixation in roots and nodules of *Alnus glutinosa*. 1. Role of phosphoenol pyruvate carboxylase and carbamyl phosphate synthetase in dark CO_2 fixation, citrulline synthesis and nitrogen fixation. *Plant Physiol.* **71**, 652–657.

Minchin, F.R. and Pate, J.S. (1973) The carbon balance of a legume and the functional economy of its nodules. *J. exp. Bot.* **24**, 259–271.

Pate, J.S. (1980) Transport and partitioning of nitrogenous solutes. *Ann. Rev. Plant Physiol.* **31**, 313–340.

Pate, J.S. and Atkins, C.A. (1983) Nitrogen uptake, transport and utilisation. In *Nitrogen Fixation* 3. *Legumes*. (ed. W.J. Broughton), Oxford University Press, Oxford, 254–298.

Pate, J.S., Atkins, C.A., White, S.T., Rainbird, R.M. and Woo, K.C. (1980) Nitrogen fixation and xylem transport in ureide-producing grain legumes *Pl. Physiol.* **65**, 961–965.

Pate, J.S. Gunning, B.E.S. and Briarty, L.G. (1969) Ultrastructure and functioning of the transport system of the leguminous root nodule. *Planta* **85**, 11–34.

Pate, J.S. Layzell, D.B. and McNeil, D.L. (1979) Modelling the transport and utilisation of carbon and nitrogen in a nodulated legume. *Plant Physiol.* **63**, 730–738.

Rainbird, R.M. and Atkins, C.A. (1981) Purification and some properties of urate oxidase from nitrogen fixing nodules of cowpea. *Biochim. Biophys. Acta* **659**, 132–140.

Raven, J.A. and Smith, F.A. (1979) Nitrogen assimilation and transport in vascular land plants in relation to intracellular pH regulation. *New Phytol.* **76**, 415–431.

Schubert, K.R. and Coker, G.T. (1981) Ammonium assimilation in *Alnus glutinosa* and *Glycine max*. *Plant Physiol.* **67**, 662–665.

Schubert, K.R. and Boland, M.J. (1984) The cellular and intracellular organisation of the reactions of ureide biogenesis in nodules of tropical legumes. In *Advances in Nitrogen Fixation Research* (eds. C. Veeger and W.E. Newton), Martinus Nijhoff, The Hague, 445–451.

Silvester, W.B. (1976) Endophyte adaptation in *Gunnera–Nostoc* symbiosis. In *Symbiotic Nitrogen Fixation in plants* (ed. P.S. Nutman), I.B.P. 7, Cambridge University Press, Cambridge, 521–538.

Thomas, R.J. and Schroder, L.E. (1981) Ureide metabolism in higher plants. *Phytochem.* **20**, 361–371.

Triplett, E.W. (1985) Intracellular nodule localisation and nodule specificity of xanthine dehydrogenase in soybean. *Plant Physiol.* **77**, 1004–1007.

Veeger, C., Haaker, H. and Laane, C. (1981) Energy transduction and nitrogen fixation. In *Current Perspectives in Nitrogen Fixation* (eds. A.H. Gibson and W.E. Newton), C.S.I.R.O, Canberra, 101–104.

Wheeler, C.T. and Lawrie, A.C. (1976) Nitrogen fixation in root nodules of alder and pea in relation to the supply of photosynthetic assimilates. In *Symbiotic Nitrogen Fixation in Plants* (ed. P.S. Nutman), I.B.P. 7, Cambridge University Press, Cambridge, 497–510.

Further Reading

Keister, D.L. and Ranga Rao, V. (1977) The physiology of acetylene reduction in pure cultures of rhizobia. In *Recent Developments in Nitrogen Fixation* (eds. W.E. Newton, J. Postgate and C. Rodriguez-Barrueco), Academic Press, London, 419–430.

Pate, J.S. (1980) Transport and partitioning of nitrogenous solutes. *Ann. Rev. Physiol.* **31**, 313–340.

Stewart, W.D.P. (1980) Some aspects of structure and fixation in N_2-fixing cyanobacteria. *Ann. Rev. Microbiol.* **34**, 497–536.

Stumpf, P.K. and Conn, E.C. (1980) *The Biochemistry of Plants* 5. *Amino Acids and Derivatives* (ed. B.J. Miflin), Academic Press, London.

Chapter 7

Alvarez-Morales, A., Dixon, R.A. and Merrick, M. (1984) Positive and negative control of the *gln* A *ntr* BC regulon in *Klebsiella pneumoniae*. *EMBO J.* **3**, 501–507.

Auger, S. and Verma, D.P.S. (1981) Induction and expression of nodule-specific host genes in effective and ineffective root nodules of soybean. *Biochemistry* **20**, 1300–1306.

Beringer, J.E. and Lazarus, C.M. (1985) Wounds activate attackers. *Nature (London)* **318**, 601.

Bisseling, T., Gover, F. Wyndaele, R., Nap, J.-P., Taanman, J.-W. and Van Kammen, A. (1984) In *Advances in Nitrogen Fixation Research*. (eds. C. Veeger and W.E. Newton). Martinus Nijhoff/Junk, The Hague, 579–586.

Cannon, F.C. and Postgate, J.R. (1976) Expression of *Klebsiella* nitrogen fixation genes, *nif*, in *Azotobacter*. *Nature (London)* **260**, 271–272.

Collins, J.J. and Brill, W.J. (1985) Control of *Klebsiella pneumoniae nif* mRNA synthesis. *J. Bacteriol.* **162**, 1186–1190.

Chua K.-Y., Pankhurst, C.E., Macdonald, P.E. Hopcraft, D.H., Jarvis, B.D.W. and Scott, D.B. (1985) Isolation and characterisation of transposon Tn5-induced symbiotic mutants of *Rhizobium loti*. *J. Bacteriol.* **162**, 335–343.

Danchin, A. and Slonimsky, P.P. (1985) Split genes. *Endeavour* **9**, 18–27.

Dixon, R.A. (1984) The genetic complexity of nitrogen fixation. *J. gen. Microbiol.* **130**, 2745–2755.

Dixon, R.A., Alvarez-Morales, A., Clements, J., Drummond, M., Filser, M. and Merrick, M. (1983) Regulation of transcription of the nitrogen fixation operons. In *Advances in Gene Technology: Molecular Genetics of Plants and Animals*, Miami Winter Symposia, Volume 20, (eds. K. Downey, R.W. Voellmy, F. Ahmad and J. Schultz), Academic Press, New York, 223–232.

Dixon, R.O.D. (1972) Hydrogenase in legume root nodule bacteroids: occurrence and properties. *Arch. Mikrobiol.* **85**, 193–201.

Downie, J.A., Ma, Q.-S., Wells, B. Knight, C.D., Hombrecher, G. and Johnston, A.W.B. (1984) The nodulation genes of *Rhizobium leguminosarum*. In *Advances in Nitrogen Fixation Research* (eds. C. Veeger and W.E. Newton)., Martinus Nijhoff/Junk, The Hague, 678.

Go, M. (1981) Correlation of DNA exonic regions with protein structural units in haemoglobin. *Nature (London)* **291**, 90–92.

Golden J.W., Robinson, S.J. and Haselkorn, R. (1985) Rearrangement of nitrogen fixing genes during heterocyst differentiation in the cyanobacterium *Anabaena*. *Nature (London)* **314**, 419–423.

Hirsch, A.M., Drake, D. Jacobs, T.W. And Long, S.R. (1985) Nodules are induced on alfalfa roots by *Agrobacterium tumifaciens* and *Rhizobium trifolii* containing small segments of the *Rhizobium meliloti* nodulation region. *J. Bacteriol.* **161**, 223–230.

Kennedy, K. and Robson, R.L. (1983) Activation of *nif* gene expression in *Azotobacter* by the *nif* A gene product of *Klebsiella pneumoniae*. *Nature (London)* **301**, 626–628.

Mclean, P.A. and Dixon, R.A. (1981) Requirement of *nif* V gene for production of wild-type nitrogenase enzyme is *Klebsiella pneumoniae*. *Nature (London)* **292**, 655–656.

Marker, K.A., Bojsen, K., Jensen, E.O., and Paludin, K. (1984) The soybean leghaemoglobin genes. In *Advances in Nitrogen Fixation Research*. (eds. C. Veeger and W.E. Newton), Martinus Nijhoff/Junk, The Hague, 573–578.

Normand, P., Downie, J.A., Johnston, A.W.B., Kieser, T. and Lalonde, M. (1985) Cloning of a multicopy plasmid from the actinorhizal nitrogen-fixing bacterium *Frankia* sp. and determination of its restriction map. *Gene* **34**, 367–370.

Nutman, P.S. (1969) Genetics of symbiosis and nitrogen fixation in legumes. *Proc. Roy. Soc.* (B) **172**, 417–438.

Nutman, P.S. (1981) Hereditary factors affecting nodulation and nitrogen fixation. In *Current Perspectives in Nitrogen Fixation* (eds. A.H. Gibson and W.E. Newton), Elsevier/North Holland, 194–204.

Quinto, C. de la, Vega, H., Flores, M. Leemans, J., Cevallos, M.A., Pardo, M.A., Azpiroz, R. Girard, M.L., Calva, E. and Palacios, R. (1985) Nitrogenase reductase: a functional multigene family in *Rhizobium phaseoli*. *Proc. nat. Acad. Sci. USA* **82**, 1170–1174.

Robson, R., Jones, R., Kennedy, C.K. Drummond, M., Ramos, J., Woodley, P.R., Wheeler, C., Chesshyre, J. and Postgate, J. (1984) Aspects of genetics of Azotobacters. In *Advances in Nitrogen Fixation Research* (eds. C. Veeger and W.E. Newton), Martinus Nijhoff/Junk, The Hague, 643–651.

Shah, V.K., Stacey, G. and Brill, W.J. (1983) Electron transport to nitrogenase. Purification and characterisation of pyruvate: flavodoxin oxidoreductase, the *nif* J gene product. *J. biol. Chem.* **258**, 12064–12068.

Smith, B.E., Dixon, R.A., Hawkes, T.R., Liang, Y.-C., Mclean, P.A. and Postgate, J.R. (1984) Nitrogenase from *nif*V mutants of *Klebsiella pneumoniae*. In *Advances in Nitrogen Fixation Research*. (Eds. C. Veeger and W.E. Newton), Martinus Nijhoff/Junk, The Hague, 139–142.

Tubb, R.S. (1974) Glutamine synthetase and ammonia regulation of nitrogenase synthesis in *Klebsiella pneumoniae*, *Nature (London)* **251**, 481–485.

Further Reading

Grierson, D. and Covey, S. (1984) *Plant Molecular Biology*. Blackie, Glasgow and London.

Verma, D.P.S. and Long, S. (1983) The molecular biology of *Rhizobium*–legume symbiosis. *Int. Rev. Cytol.* Suppl. **14**, 211–245.

Chapter 8

Akkermans A.D.L. and Van Dijk, C. (1981) Non-leguminous root nodule symbioses with actinomycetes and *Rhizobium*. In *Nitrogen Fixation* 1: *Ecology* (ed. W.J. Broughton), Clarendon Press, Oxford, 57–103.

Allen, E.K. and Allen, O.N. (1958) Biological aspects of nitrogen fixation. *Encyclopedia of Plant Physiology*, Vol. 8, (ed. W. Ruhland), Springer-Verlag, Berlin, 48–105.

Armstrong, F. (1985) Cyclic fixation of nitrogen. *Nature (London)* **317**, 576–577.

ASA (1976) *Multiple Cropping*. American Society of Agronomy Special Publication 27, American Society of Agronomy, Madison, Wisconsin.

Becking, J.H. (1984) Identification of the endophyte of *Dryas* and *Rubus* (Rosaceae). *Pl. Soil* **78**, 105–128.

Burton, J.C. (1981) New developments in inoculating legumes. In *Recent Advances in Biological Nitrogen Fixation* (ed. N.S. Subba Rao), Edward Arnold, London, 380–405.

Burris, R.H. (1977) A synthesis paper on non-leguminous H_2 fixing systems. In *Recent Developments in Nitrogen Fixation* (ed. W.E. Newton, J.R. Postgate and C. Rodriguez Barrueco), Academic Press, London, 487–511.

Ciba Foundation (1983) *Better Crops for Food*. Symposium 97, Pitman, London.

Coucouvanis, D. (1984) The synthesis, structures and electronic properties of Fe–Mo–S polynuclear aggregates. Molecules of elementary structural compliance with the Fe–Mo–S aggregate of nitrogenase. In *Advances in Nitrogen Fixation Research* (eds. C. Veeger and W.E. Newton), Martinus Nijhoff, The Hague, 81–91.

Daft, M.J. Clelland, D.M. and Gardner, I.C. (1985) Symbiosis with endomycorrhizas and nitrogen-fixing organisms. *Proc. Roy. Soc. Edin.* **85**(B) 283–298.

Dale, J.E. (1979) Nitrogen supply and utilization in relation to development of the cereal seeding. In *Nitrogen Assimilation of Plants* (eds. E.J. Hewitt and C.V. Cutting), Academic Press, London, 153–163.

Dandekar, A.M., Lerudulier, D., Smith, L.T., Jakowec, M.W., Gong, L.S. and Valentine, R.C. (1983) In *Plant Molecular Biology* (ed. R.B. Goldberg), Alan R. Liss, New York, 277–289.

Dixon, R.A. and Postgate, J.R. (1972) Genetic transfer of nitrogen-fixation from *Klebsiella pneumoniae* to *Escherichia coli*.. *Nature* **237**, 102–103.
Franco, A.A. (1979) Contribution of the legume–*Rhizobium* symbiosis to the ecosystem and food production. In *Limitations and Potentials for Biological Nitrogen Fixation in the Tropics* (eds. J. Dobereiner, R.H. Burris and A. Hollaender), Plenum Press, New York 65–74.
Gauthier, D. Diem, H.G. and Dommergues, Y.R. (1985) Assessment of N_2 fixation by *Casuarina equisetifolia* inoculated with *Frankia* ORS 021001 using ^{15}N methods. *Soil. Biol. Biochem.* **17**, 375–379.
Gisby, P.E. and Hall, D.O. (1980) Biophotolytic H_2 production using alginate-immobilised chloroplasts, enzymes and synthetic catalysts. *Nature (London)* **287**, 251–253.
Gordon, J.C. and Wheeler, C.T. (eds.) (1983) *Biological Nitrogen Fixation in Forest Ecosystems: Foundations and Applications*. Martinus Nijhoff, The Hague.
Grierson, D. and Covey, S. (1984) *Plant Molecular Biology*. Blackie, Glasgow and London.
Harley, J.L. and Smith, S.E. (1983) *Mycorrhizal Symbiosis*. Academic Press, London.
Knowlton, S., Berry, A. and Torrey, J.G. (1980) Evidence that associated soil bacteria may influence root hair infection of actinorhizal plants by *Frankia*. *Can. J. Microbiol.* **26**, 971–977.
Knowlton, S. and Dawson, J.O. (1983) Effects of *Pseudomonas cepacia* and cultural factors on the nodulation of *Alnus rubra* roots by *Frankia Can. J. Bot.* **61**, 2887–2892.
Leaver, C.J. and Gray, M.W. (1982) Mitochondrial genome organisation and expression in higher plants. *Ann. Rev. Plant Physiol.* **33**, 373–402.
Mugnier, G.J.J., Diem, H.G. and Dommergues, Y.R. (1982) Polymer-entrapped *Rhizobium* as an inoculant for legumes. *Plant Soil* **65**, 219–231.
Pickett, C.J. and Talarmin, J. (1985) Electrosynthesis of ammonia. *Nature* **317**, 652–653.
Stewart, W.D.P., Sampaio, M.J., Isichei, A.O. and Sylvester-Bradley, R. (1979) Nitrogen fixation by soil algae of temperate and tropical soils. In *Limitations and Potentials for Biological Nitrogen Fixation in the Tropics* (eds. J. Dobereiner, R.H. Burris and A. Hollaender), Plenum Press, New York, 41–63.
Steyaert, R.L. (1932) Une épiphyte bactérienne des racines de *Coffea robusta* et *C. klainii*. *Rev. Zool. Bot. Afr.* **XXII**, 133–139.
Talley, S.N. and Talley, B.J. (1978) Nitrogen fixation by *Azolla* in rice fields. In *Genetic Engineering for Nitrogen Fixation* (ed. A. Hollaender), Plenum Press, New York, 259–281.
Tarrant, R.F. (1983) Nitrogen fixation in North American forestry: Research and applications. In *Biological Nitrogen Fixation in Temperate Forests; Foundations and Applications* (eds. J.C. Gordon and C.T. Wheeler), Martinus Nijhoff, The Hague, 261–278.

Further Reading

Harley, J.L. and Smith, S.E. (1983) *Mycorrhizal Symbiosis*. Academic Press, London.
Normand, P. and Lalonde, M. (1986) The genetics of *Frankia*: a review. *Plant & Soil* **90**, 429–435.
Subba Rao, N.S. (1984) Interaction of nitrogen-fixing microorganisms with other soil microorganisms. In *Current Developments in Biological Nitrogen Fixation* (ed. N.S. Subba Rao) Pitman, London, 37–64.

Index

acetylene reduction 58, 76
actinorhizal root nodules 25 *et seq.*
  ammonium effects on 92
  carbon reserves in 85, 86
  development of 28
  gaseous diffusion in 75–80
  haemoglobin in 34, 78, 81, 82
  homogenates of 75
  hydrogen evolution by 80, 84
  nodule structure of 105
  perennation in 84–6
  solute transport in 105, 106
  sporangia of 30
  spores of 84
  vesicles in 30, 76, 77, 80, 90, 91, 93
Aeschynomene afraspera 11
Agrobacterium rhizogenes 9
  rubi 9
  tumefaciens 9, 119, 140–2
agroforestry 130
Albizzia 96
alder 3
allantoic acid 96–100, 104
allantoin 96–100, 104
Alnus 29, 31, 76, 81, 82, 90, 91, 96, 100, 126
Alnus crispa 78, 80
  glutinosa 84, 92
  incana 92
  rugosa 84
  rubra 25, 85
amides 93, 94, 96, 104–6
ammonia 113
ammonium
  nodulation and 91, 92
  ureide synthesis and 106
  assimilation of 92–102
  membrane potential and 89, 90
  mRNA stability and 88
  nitrogenase regulation by 87–91, 113
Anabaena 102, 103, 117
Anabaena azollae 39
  cylindrica 90
anaerobes 3

Anthoceros 102
antibiotic resistance 137
Arachis hypogea 35
archaebacteria 3
arginine 111, 114
asparagine 74, 93–5, 97, 104, 106
aspartic acid 74
ATP
  amide biosynthesis and 92–5
  citrulline synthesis and 100, 101
  nitrogen fixation, nitrate reduction and 133
  nitrogenase activity and 47, 89
  ureide synthesis and 97–100
aut gene 111
Azolla 3, 37–40
  structure of 39, 102
  transfer of solutes in 39, 102
Azolla caroliniana 102
  filiculoides 134
Azospirillum 8
Azotobacter 6, 54, 56, 134
  ammonium effects on 97–9
  genetics of 115–6
Azotobacter paspali 8

bacteroids 22
Blasia 102
bleeding xylem sap 104–6
Bradyrhizobium 9

carbamyl phosphate 84, 100, 101
carbon dioxide
  diffusion of 63, 77, 80
  fixation of 63, 84, 100
carbon monoxide inhibition 109
carbonic anhydrase 63, 80
Casuarina 32, 76, 81, 128
Casuarina cunninghamiana 81, 82
  equisetifolia 76, 126, 128, 129
cDNA 122
Ceanothus 32
cephalodia 36
citric acid cycle 74

INDEX

citrulline 84, 96, 128, 129
Clostridium 5, 54
Coffea 131
Colletia 32
component 1 protein 49
component 2 protein 49
Comptonia 32
Comptonia peregrina 11, 81, 82, 84
conformational protection 7
Coriaria 32
cyanobacteria 3, 13, 35–44
  in *Azolla* 37–40, 102
  in cycads 40–2, 102, 103
  in *Gunnera* 42–4
  in lichens 36
  in liverworts 102
Cycadaceae 40

Datisca 32
Derxia gummosa 8
Desulfotomaculum 5
Desulphovibrio 5
determinate nodules 20
Discaria 32
Dryas 32

effective nodules 24
Elaeagnus 27, 81
electron transport 54, 109
energy requirements
  for amide biosynthesis 95
  for citrulline synthesis 100, 101
  for nitrate reduction 133
  for ureide synthesis 98, 100
Erwinia herbicola 132
Escherichia coli 132, 139
exons 123

facultative anaerobes 5
Fe protein 49, 109
Fe–S cluster 109
FeMoco 47, 50, 109, 143
ferredoxin 47, 54
fertilizer manufacture 45, 142, 143
FeSII 7
fixation thread 35
flavodoxin 54, 109
Frankia 11–12, 27 *et seq.*, 56
  acetylene reduction by 75–86, 126, 137
  ammonium effects on 77
  carbon metabolism in 81–84
  cell sugar patterns of 81
  gaseous diffusion in 77, 78
  hydrogenase in 80
  isolation of 75
  nitrogenase in 76, 77
  oxygen in 76–8
  respiration of 77, 83
  strain classification of 81–3
  vesicle structure in 76, 77, 80, 84, 90, 91, 93

gaseous diffusion 61–72, 75–80
genetic engineering 131–42
genetics 107 *et seq.*
Gloeotheca 14
gln A gene 111
glutamate synthase (GOGAT) 93, 94, 102
glutamic dehydrogenase 93, 100, 102, 111
glutamine synthesis 93–5, 103, 113
  in alder nodules 100
  in *Azolla* 102
  in lichens 102
  reactions of 92–4
  regulation of nitrogenase in 87, 88, 91
glutamine synthetase 37, 40, 111, 113
glycine 96
glycolytic pathway 72
Gunnera 42, 103

Hac phenotype 120
haem synthesis 119
haemoglobin 34, 75, 81, 82
helper bacteria 127
heterocysts 13, 36, 37, 40, 42, 77, 90
Hippophae 32
his gene 108
histidine 111, 114
host genes 120
hup gene 118
hydrogen 70
  and ammonia synthesis 143
  evolution of in actinorhizal nodules 80, 84
  evolution of in legume nodules 55–7
hydrogenase 8, 71
  in actinorhizal nodules 80
  in legume nodules 55–7

INDEX

indeterminate nodules 20
ineffective nodules 25
infection
  in actinorhizal plants 27
  in legumes 18
infection thread 20, 22, 35
inoculation procedures 126, 127
inosine monophosphate 96, 97
intercellular spaces 64
intercropping 128
introns 123
iron–sulphur clusters 47–8

Kjeldahl method 57
Klebsiella 5, 87, 88, 132, 139
  genetics of 110–12, 115

lectins 16–17, 37
leghaemoglobin 21, 23, 62–70, 121, 123
  function of 68
  genes for 121, 123
  location of 68
  synthesis of 66
legume nodules
  and carbon dioxide 62
  and transpiration 105
  bacteroids in 22
  biochemistry of 61 *et seq.*
  determinate 20
  development of 20
  effective 24
  gaseous exchange in 61
  hydrogen evolution by 70
  hydrogenase 55–7
  indeterminate 20
  ineffective 25
  infection of 18–25
  infection thread in 20, 22
  leghaemoglobin in 21, 23, 63
  nitrogen assimilation by 92–100
  nitrogen export by 103–6
  nitrogenase in 49–50
  oxygen in 64–70
  peribacteroid membrane in 22
  physiology of 61 *et seq.*
  senescence of 91
  structure of 98–100
  transfer cells in 22
  vascular systems of 103–5
Leguminosae 15
Leucaena leucephala 128, 129

lichens 36
  ammonium assimilation in 102
Lotus 116
Lotus pedunculatus 119
Lupinus 96
Lupinus luteus 67

Macrozamia 103
Macrozamia communis 40
Methanococcus
  thermolithotropus 4
Methanosarcina barkeri 4
methionine sulphoxime 87, 91
mitochondria 133, 134
Mo–Fe–S cluster 50
MoFe protein 49–50, 108–9
molybdenum cofactor 50, 108, 109
mRNA 88
mutagenesis 135–137
mycorrhiza 127
Myrica 76, 80, 82
Myrica gale 30, 31, 81, 84
  pensylvanica 80

NADP(H) 55
nif genes 5, 107 *et seq.*
  mutagenesis of 135–7
  regulation of 110
  repression of 87, 88, 91
  transfer of 132–5
nitrate
  and nodulation 91
  nitrogenase repression by 88, 90, 91
  reduction, energy requirement for 132
nitrite 88, 90, 91, 106
nitrogen 45
nitrogen fixation 45, 133
  annual rates of 126
  applications of 125–30
  energy requirements of 91, 133
  industrial 142, 143
  and photosynthesis 91
  respiratory costs of 84
  seasonality of 86
nitrogenase 49 *et seq.*
  and membrane potential 88–91
  and ATP 89
  and combined nitrogen 87–92
  and MgATP 51
  ATP requirement of 47
  covalent modification of 90
  Frankia and 76, 77

hydrogen evolution by 53, 56
in *Frankia* 76, 77
in photosynthetic bacteria 90
location of 30
mechanism of action of 51
oxygen and 5
reductant 47
regulation of by ammonium 87–91
regulation of by glutamine 87, 88, 91
nitrogenase substrates of 52
temperature–sensitivity of 137
nod genes 124
nodulation
 and combined nitrogen 91
 in new species 130, 131
nodules, water stress in 138
nodulins 121–3
Noi phenotype 120
non-haem iron protein 47
Nostoc 40, 42, 102
ntr genes 111

operon 111, 112
opines 140, 142
Ornithopus sativus 67
oxygen
 and actinorhizal nodules 75–81
 and legume nodules 64–70

Parasponia 11, 27, 34, 131
 and gaseous diffusion 75
 nodule structure in 76
Paspalum notatum 8
Peltigera aphthosa 36, 102
Peltigera canina 36
PEP carboxylase 3, 74, 84
peribacteroid membrane 22
Phaseoleae 107
Phaseolus 96
phosphoenol pyruvate 63
phosphoroclastic reaction 46, 54
phytohaemagglutinin 16
Pisum 96
Pisum sativum 81, 82
plasmid 10, 115, 119, 132, 140–2
proline 111, 114, 138
promoter 117
propionyl CoA carboxylase 82, 83
protein A 123
Proteus mirabilis 123

protoplasts 142
Pseudomonas cepacia 28
purine synthesis 96
put gene 111
pyruvate 47
pyruvate flavodoxin oxido-reductase 110

Relative Efficiency 70
respiratory protection 6
Rhizobium 9–11, 56
 ammonium effects on 90
 ammonium excretion by 90, 99
 extending host range of 137
 genes in 116
 inoculation procedures for 126
 strain competition in 137
Rhizobium japonicum 54, 68, 116
 leguminosarum 116, 118, 119
 loti 119
 meliloti 116, 132
 phaseoli 117
Rhodopseudomonas 13, 117
Rhodospirillaceae 90
Rhodospirillum 13
Rubus 32
Rubus ellipticus 131

Salmonella typhimurium 132
Serratia marcescens 132
Sesbiana rostrata 10, 91
Shi A gene 108
shikimic acid 108
site-directed mutagenesis 135
Spirillum lipoferum 126
sporangia 30
Staphylococcus aureus 123
stem nodules 10, 91
Sterocaulon vesuvianum 36
sym plasmid 119
symbiotic genes 119

Ti plasmid 140
transfer cells 22, 39, 103
transfer of fixed nitrogen 93, 96, 101–6
transposon 120
Trifolium pratense 120

uptake hydrogenase 8, 56, 71, 80, 118

ureides 95–101, 103–6
uricase 98

vesicles 30
Vicia 96
Vicia faba 118

Vigna 96
virus transfer 142

water stress 138, 139

xylem sap 93, 104, 106

THE LIBRARY
ST. MARY'S COLLEGE OF MARYLAND
ST. MARY'S CITY, MARYLAND 20686